THE AMAZING CRAWFISH BOAT

Folklore Studies in a Multicultural World

The Folklore Studies in a Multicultural World series is a collaborative venture of the University of Illinois Press, the University Press of Mississippi, the University of Wisconsin Press, and the American Folklore Society, made possible by a generous grant from the Andrew W. Mellon Foundation. The series emphasizes the interdisciplinary and international nature of current folklore scholarship, documenting connections between communities and their cultural production. Series volumes highlights aspects of folklore studies such as world folk cultures, folk art and music, foodways, dance, African American and ethnic studies, gender and queer studies, and popular culture.

THE AMAZING CRAWFISH BOAT

JOHN LAUDUN

University Press of Mississippi   Jackson

www.upress.state.ms.us

The University Press of Mississippi is a member of the Association of American University Presses.

All illustrations are by the author unless otherwise noted.

Publication of this book is supported by a grant from the Andrew W. Mellon Foundation.

First printing 2016

∞

Library of Congress Cataloging-in-Publication Data

Names: Laudun, John, author.
Title: The amazing crawfish boat / John Laudun.
Description: Jackson, Mississippi : University Press of Mississippi, 2016. |
   Series: Folklore studies in a multicultural world series | Includes
   bibliographical references and index.
Identifiers: LCCN 2015025349 | ISBN 9781496804204 (hardback) | ISBN
   9781496804211 (ebook)
Subjects: LCSH: Crayfish culture—Louisiana—History. |
   Crayfish—ouisiana—History. | BISAC: SOCIAL SCIENCE / Folklore &
   Mythology. | TECHNOLOGY & ENGINEERING / Fisheries & Aquaculture.
Classification: LCC SH380.92.U55 L38 2016 | DDC 639/.6409763—dc23 LC record
available at http://lccn.loc.gov/2015025349

British Library Cataloging-in-Publication Data available

FOR MY PARENTS—
who, through an endless parade of gadgets,
each with its own fierce optimism for a better world,
made me into the curious optimist that I am.

"How do I know? Experience, sir, experience."—Dale Olinger

CONTENTS

~~~~~~~~~~~~~~~~

INTRODUCTION

This is the story of the crawfish boat, a very particular machine that seems to have emerged on a particular landscape at a particular point in time. Because of these particularities, it might be tempting to see the story as confined, constrained in meaning and relevancy, but in order to understand how the boat came to be, we have to understand the landscape out of which it emerged. In order to understand the landscape, we have to see it as the people who live and work on it see it. In order to understand them, we need to listen to what they say and watch what they do, and that is never as easy as it sounds. But if there is science in this, it is in the particularities, the exact facts that are to be found both in the things that we can see and touch as well as the things that we can hear if we linger long enough in doorways as people tell us about the world as they understand it.

And so this is really the story about the people who brought the crawfish boat into existence, almost whole-cloth, like Athena springing from Zeus's brow. In telling this story, I am both reaching back into the past as well as recounting events that happened only yesterday and will happen again tomorrow. The men who invented the crawfish boat were and are farmers and fabricators. They are ordinary men: they get up in the morning and walk or drive to their shops or equipment sheds, which in many instances are next door to their homes.

If they are farmers, they survey fields full of rice or soybeans or full of water for crawfish. They worry about the weather. Will it rain in time so that they do not need to start up their well pumps or has it rained so much that they need to pull drains and let things dry? Will the predicted severe storms lay their rice crop, tan and heavy with full kernels, down on the ground, making it impossible to harvest, or will they have the chance to run their combines, harvest their rice, and take their chances with what the mills are paying this year? How they answer those questions will determine what tractor they climb into, what tool they attach to it, and how they will proceed throughout their day. They will work mostly alone, on no one's clock but their own in consultation with Mother Nature, whose indulgence they court and whose scorn they consider their fitful burden.

If they are fabricators, they are already thinking about the jobs queued up for the day, fully knowing that at least one major and several minor emergencies will come barging into the shop to break up any neatly planned schedule. A farmer will need something tacked back together well enough to work for the time being and they will need it now. Someone else will come in with a job and they will want to discuss it in detail now. Some piece of inventory that they thought they had in stock and is required for a job will need to be ordered now, and it will require cleaning hands and going into the office where they will be chilled by the lovely air-conditioning and comforted by a soft desk chair only to have to rise again to crawl back under a tractor or grain cart or crawfish boat, reaching past crusted mud or rotting rice or crawfish or hanging wasp nests, in order to get back to work.

They are ordinary men. They are, as the historian Francis Andrews observed about makers from another time and place, men who are "doing ordinary work as they know it . . . [without] any idea of doing a great and notable thing or one privately inspired by any name to be made in the success of it."[1] What we know about the predecessors of our modern fabricators, the medieval stonemasons who built the great Gothic cathedrals, we have gleaned from occasional mentions in the historical record. The architect John James lamented about these medieval artisans that "little is known about the builders. A name crops up here and there, and a few comments assure us that they were no more anonymous than we are. We read of [them] sitting at the high table with their clients." The mentions are few, however, and we are faced with the fact that very little is known about the actual men who created Gothic architecture: "One of the world's great art styles would seem to have come into being without leaving a trace of the people who made it."[2]

One response is not to care, to write history as if ages produce buildings or cultures invent jokes. It is not a terrible shorthand. It is not a whole-cloth obfuscation of any underlying actualities. But such abstractions, sometimes idealizations, do shorten and obscure a much more complex reality that deserves our full attention, if only for a short time; a reality that is complex because it is made up of people, people with different attitudes and different experiences.

Andrews tried his best to chronicle and understand medieval builders, men he could not know because of the great gulf that is time. We are fortunate with the crawfish boat. Its history is relatively recent, and most of the men who had a hand in its development are very much alive. We can talk with them and ask them what they did. They may not remember very much—one day twenty or thirty years ago often blurs into another day twenty or thirty years ago—or very well, but if we talk to enough people, we can begin to puzzle together where the idea seems to have originated and how it developed.

Some may wonder, why bother with the crawfish boat at all? It's no Gothic cathedral.[3] It is a smallish object with a small impact. Its geography is focused mostly in southwest Louisiana and those parts of Texas and Arkansas where rice is also grown. Its invention would seem a happy historical happenstance, and its continued production locally a function of a market too small to be noticed by bigger firms, firms who employ proper engineers and research scientists who work with CAD/CAM systems and elaborate workflow systems that produce reams of documentation that facilitate those historians who sometimes traipse after them to record the innovations they make and thus to certify their ingenuity.

The boat makers themselves wonder much the same thing: what's the big deal? They saw a problem, and they came up with a solution. They did it in collaboration with family, friends, and acquaintances. They did it through observing what worked and offering their own version of the steadily evolving adaptation that was the boat itself. Some of their adaptations were taken up by others, some not. They are simply businessmen or farmers trying to make a living: the boat offered them a way to make money. By and large, they do not consider themselves inventors; rather, among themselves, they are repairmen, fabricators, welders, and/or farmers, with each man possessing some or all of the skills of these roles to varying degrees. Again, from their point of view, any abstraction that describes their work reaches no further than this.

What they imagine they have done is figured out how to profitably get crawfish out of shallow fields flooded for the purpose and into the backs of pickup trucks that shuttle the thirty-pound sacks of living creatures to processing plants or restaurants in such a timely fashion and on such a scale that everyone involved can make some money.

In answering that economic riddle, however, they have managed to make a kind of aquaculture arise on a landscape where once there was only agriculture. It's amazing, really. Two centuries ago, this was a pastoral landscape, filled with herds of cattle that were moved to regional processing centers and then shipped off to cities. A century ago, rice began to displace cattle on the Louisiana prairies, and now rice fields are regularly turned into ponds with neat lines of crawfish traps serpentining through them.

In a scant century or so, this prairie landscape has changed from pasture to agriculture to aquaculture, with all three now coexisting side by side. Perhaps history is inured to the individuals, and their discrete actions, that made that change possible, but the magnitude of the change, seen readily even in satellite imagery, can hardly be ignored: the boats leave telltale signs of their paths in most fields.

So it was a boat that changed farming in the region—I have heard some men jokingly refer to themselves as *crawfish farmers*—and everyone calls them boats. They have hulls. They float. But beyond these simple facts, they don't look like any boat most of us have ever seen or will ever see. The entire craft runs off what appears to be an oversized lawn-mower engine. The engine drives a pump, and the pump drives a hydraulic motor, or set of motors, that powers a great steel wheel that hangs off the back of the boat. The wheel itself looks like something out of a bygone era: its size and shape and its cleats, as they are called, are reminiscent of steamboat paddle wheels. But the cleats aren't paddles pushing water: they are treads designed to give the wheel traction as it rolls along the bottom of a flooded field. The advantage this offers over paddles or propellers is that the craft stops instantly and turns quickly, unlike the gliding halt and slower turns of a fully aquatic boat. The other advantage of the great wheel is what still turns heads: the sight of a boat driving down a road—in fact, this is such an important part of a crawfish boat that one maker puts extra cleats on his wheels so they roll smoothly.

How did this happen? Who brought this machine into existence? Who continues to contribute to its production and refinement?

There is little doubt that right now the dominant maker, by sheer force of numbers, is Kurt Venable of Rayne, who has focused a good part of his fabrication shop's business on the crawfish boat. Next, in terms of volume, is Mike Richard, who operates a one-man aluminum welding business just south of Eunice. Making fewer boats than he once did but still a significant

Crawfish boat routes through a collection of rice fields that lie along Interstate 10 west of Rayne. Courtesy of the Aerial Photography Field Office, USDA Farm Service Agency.

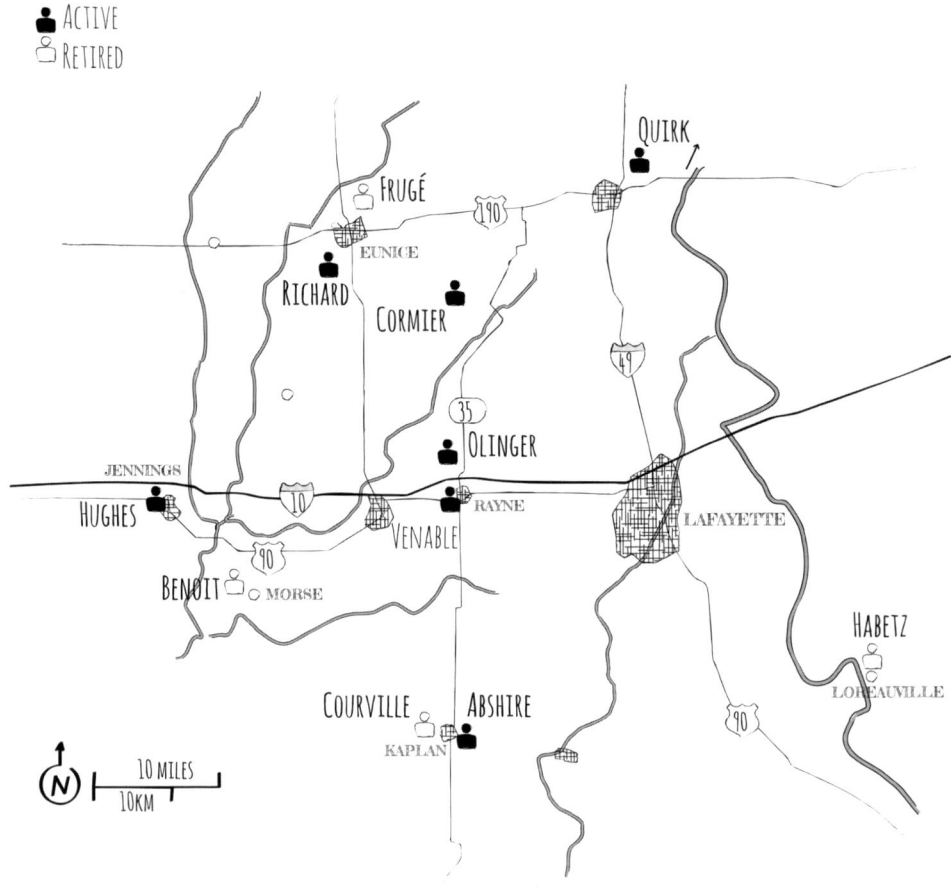

Map of Louisiana showing all the makers who have contributed, or continue to contribute, to the development of the crawfish boat.

part of the network is Gerard Olinger, who runs an agricultural-equipment repair shop north of Rayne in the German settlement of Roberts Cove.

In addition to these makers, there is Dale Hughes, who runs a fabrication business in Jennings; Michael Quirk in Lebeau; and Mike Cormier, a farmer outside Church Point who occasionally builds an extra boat or two. In Kaplan, there was Clayton Courville, and there remains Jimmy Abshire and his brother Robert, who operate a repair shop and equipment-supply store. And in the past, there was Tedmon Habetz, the man widely credited with creating the first hydraulic crawfish boat, and Harold Benoit, who helped organize the first crawfish field day at which Habetz's boat premiered, and who had also built a hydraulically powered boat but had not yet figured out how to slow it down enough to make it workable. Finally, there was Greg Frugé, the man who seems to have made the crawfish boat popular and is affectionately known among many of the makers and farmers of his generation as "Momma Greg." Working in Eunice, far from Habetz in Loreauville and Benoit in Morse, Frugé developed a rather interesting drive unit that was in fact based on the mechanical transfer of power, but he eventually built hydraulic boats as well.

Together, these makers are part of a dynamic, diffuse network. Each man works (or worked) alone, separate from the other builders. At the same time, he is also working in dialogue with them. They see each other's boats. They know what their customers think of the strengths and weaknesses of each design. They recognize imagination in each other, and, in that way, share their creativity. It's not unlike a guitar player hearing a run or riff, liking it, and including a version of it in his or her next performance. Listening, even indirectly, is what creates an artistic field. Out of that field comes creativity, as artists and craftsmen spur each other on.

Such an idea is terribly important, of course, given current concerns about the state of creativity, especially in relation to industry, in the United States, but perhaps it can be said quite clearly up front, allowing time then for the unpacking across the rest of this exploration: Although I began my journey with the intent of discovering how residents of Louisiana understand the landscape on which they live and work, what I found was a creative engine that seems always to have been present on that landscape and that, I think, throws into stark relief the nature of the relationship between tradition and creativity, two terms we too often place at opposite ends of a spectrum of activity. Such a discovery is not new, of course. Others have come before me in this regard, but it bears repeating that what is revealed here is a cultural background more varied than most imagine that underlies a network of makers—of craftsmen—who are quite varied in their own natures.[4]

Currently, the study of creativity is a vast enterprise. Within it, there must be a place for individuals whose eminence is bounded by locale, either by preference or by providence. Such individuals give us a glimpse of the nature of the creative act in an immediate and intimate fashion. That is, confined to a definable horizon, the creative act reveals the competence of the individual in the very moment of performance. Folklorists have long studied creativity, even if we were sometimes discerning its shape by its shadow, alongside other humanists, but in the past few decades we have been joined by an increasing number of scientists who, whether interested in the mechanics of the brain or in the way markets respond to novelty, work under the collective umbrella of creativity studies. A few initial forays into embedded, or contextual, studies of creativity—labeled *case studies*—within the larger field have been assayed, but it is early in their development, and I believe folklore studies stands to make a ready contribution to their efforts, offering as we can our decades-long refinement of the ethnographic study of creative moments.[5]

The trick, of course, is to study human beings as they are, always caught between being "free and stuck in the world," as Henry Glassie noted.[6] Absolute freedom is where the humanities have tended to focus their attention: on artists who, alone in their studios or garrets, are able to explore the furthest reaches of what is possible to imagine and then realize it in some fashion without concern for audiences or markets. At the other end of the spectrum are those who, we imagine, are so stuck within the confines of everyday existence that they cannot see anything else, let alone accept any novelty, whether it be intentional or random.

What Glassie found among the farmers of Ballymenone in Ireland is what Charles Zug found among the potters of North Carolina, and what Janet Gilmore found among the boatbuilders of Coos Bay, Oregon: there is always a diversity of players within a scene, some of which is due to talent—itself really a function of drive and inclination more than anything else—and some of which is due to what the local ecology can bear: there is room only for so many of a certain kind of specialist within a local economy.[7] It was this kind of dense network of people and ideas that beckoned me when I began my own study of creativity on a different landscape, one filled with water and thus requiring a special machine to traverse it. I knew, too, that I wanted to address directly the antipodal anchors of creativity studies, the starry-eyed dreamer or the bloody-eyed laborer, and so I found myself drawn to an extraordinary artifact whose very realization screamed creativity and yet whose natal scene was grimy, noisy, and as modern as one could imagine it.

The south Louisiana landscape is dotted with shops where metal is either bent to serve new purposes or unbent to serve a familiar purpose again. Some of the shops serve tasks familiar to modernity everywhere: the thin galvanized sheets that are folded into air-conditioning ductwork or the curved panels of our automobiles that need to be sleek again after a mishap. Common to rural landscapes are the shops that repair or fabricate agricultural equipment. Common to industrial landscapes are shops that build heavy-duty structures for factories or, in south Louisiana, drilling rigs. Some of those shops line the waterways of the region, giving themselves easy access to a transportation network that can carry heavy, wide loads out to the Gulf of Mexico. Those shops are joined by others that specialize in making a variety of watercraft that serve the oil industry, among others.

The result is a dense network of shops where those men, and some women, who think best in three-dimensional shapes, in mathematical ratios, and in stress factors find an outlet for their imagination. There is room, as there always is, for individuals who simply want to be told what to do, but there is also room for individuals who want to excel. And if there isn't in one shop, there will be another, or if there is none to be had in the present moment, then some will open shops of their own. That's how Kurt Venable and Mike Richard got their starts.[8]

Most of these men work alone or with a small collection of trusted others. Often this group is composed of members of their family: brothers, wives, sons. Each man works on that combination of things that interests him and also pays the bills. Venable likes to design workflows. Richard likes anything aluminum. Olinger likes difficult repair jobs. So a landscape filled with a seemingly repetitive series of shops, each possessed of various congregations of men with grease under their fingernails, is actually quite varied.

There are only a few manufacturing secrets here and there that each man possesses because everything there is to know is in a boat. Every hard-won idea must manifest itself in steel or aluminum where it is available for all to see, analyze, and judge. And there is almost no end to the discussion of who makes a better boat or whose boat is best suited for which soil or terrain type. The makers themselves are judged for the quality of their boats, their willingness to customize a boat, and their willingness to repair or modify a boat made by someone else.

A BRIEF NOTE

A quick word about the terms and methods used in the research and writing of this text. The interviews and observations that inform the writing took place over the course of six years. When it was possible to take careful notes, I took them, but quite often I was standing in the middle of a metal shop or riding on a jump seat of a tractor or standing at the edge of a hot field on a summer day in Louisiana, and taking notes in those moments was quite difficult, as the many pages in my notebooks smeared with grease and mud or wrinkled from dripping sweat can attest. I am fairly confident that in every case the words attributed to individuals represented herein are as exact as my note-taking or memory could make them. Any slippage is entirely my responsibility.

Where conversations were recorded and their results transcribed, I have hewn to a more novelistic way of representing speech. In other contexts, I have engaged in more elaborate forms of transcription, for the sake of understanding the nature of discourse and the flow of discursive interaction. Such was not my goal here, and invoking that kind of involved typographic apparatus here seemed inappropriate. For those interested more particularly in the how of speaking of the individuals represented here, I plan to deposit the recordings, both audio and the thousands upon thousands of images, with an appropriate archive, much as I have made some video recordings available through the EVIA (Ethnographic Video for Instruction and Analysis) Digital Archive.

As I repeatedly remind institutional boards that worry about the destruction of sensitive data, as a folklorist it is my job to work with the people whom I study to create a historical record. The last thing we, they and I, want is to destroy records. In many cases, some of the individuals, or their acts chronicled here, will too soon be lost to history. If for no one else but their children, grandchildren, neighbors, and friends, my hope is that a text like this can not only preserve a memory or two but also act as a powerful reminder that each of us, just like the individuals in these pages, has the ability to make something no one has seen before. And even once it is fully formed, we can, using our own particular collection of experiences and expertise—and dare I add obsessions?—innovate within that form and add value, a phrase that has become something of a cliché in our contemporary landscape. We can reclaim the cliché, however, by insisting on the place of people in all of this and for the value to be not only economic but also social.

The importance of people as people is something on which my discipline, folklore studies, prides itself. Not people as aggregated into numbers. Not people as broken into traits. People as people, as individuals who are both collections of traits and pieces of a larger whole at the same time. My good friend Henry Glassie once noted that folklore studies is indeed a science, but a science that treats humans as humans and not as monkeys.[9] Let me be clear, I am not opposed to science. Far from it. The objectification of complex realities is often the first step in understanding them. But, and here is where the art lies in science, deciding upon the nature of the object, what it is, is not something that should be rushed. The annals of applied science and social science are awash in stories of well-intentioned studies or efforts gone astray because they failed to take time to understand the larger ecosystem. The current study is but one small step toward understanding the nature of creativity within a complex system of ideas, actors, and events. There is so much more work to be done in this regard.

Finally, I would be remiss if I did not mention that readers may very well note the overwhelming use of the masculine pronoun, even when in reference to abstract entities. The truth is, all of the makers surveyed here are men. There are a number of women who play a variety of roles, including operators of the boats, wives who most assuredly had some input at various moments, and, most especially in the case of Sheryl Venable, actual managers of the businesses. But the people who make these boats in the present, and those remembered in the past, are all men. It made the most sense, then, simply to revert to the use of masculine pronouns in the narration and discussion of the men involved. The larger role of women in the everyday working of farms and shops, as well as the role of African Americans and Mexican farm laborers, will have to await a more comprehensive ethnographic study. This text is not that, although I have certainly begun to understand all the things scholars of Louisiana culture and history have yet to document, analyze, and understand. More on that at the end of this book.

THE AMAZING CRAWFISH BOAT

Randy Gossen swinging a crawfish trap above his boat's sorting table.

FORWARDS

The wind that blew lightly over the freshly plowed rice fields was just cold enough to chill exposed fingers and cheeks and just strong enough to rustle nearby trees. Perched somewhere unseen, a few birds whistled their wakefulness. Moving across a narrow blacktop road, the wind picked at the surface of a flooded field, transforming its smooth surface into thousands of fractals, each reflecting the sun, still low in the sky.

It was, for all the world, a quiet country morning until a small engine clattered to life. Its owner ran it up for a moment, and then let it settle down to an idle that would warm it to the day's work. The sound of the motor was high, almost nasal when compared to the throatier roar of the big diesel engines that typically make their way across these fields powering tractors. This sound was more like something you would hear on a suburban lawn than an agricultural field. And that was about right, since this motor could only offer up twenty-five horsepower.

The motor continued its fierce vibrato while Randy Gossen finished loading his boat with bait for the morning's run. The plastic tubs were full of frozen fish chopped, depending upon their original size, in halves and thirds. The fish were chub, trash fish to most fishermen and a bycatch of the Louisiana menhaden fishery. The tubs were the same ones seen in any retail store when shelves are being restocked. The chopped chub were packed tightly in the tub, but they still managed to shift a bit as Gossen slid them off the lowered tailgate of his truck and onto the floor of the boat. He countered the shift using his tall frame to his advantage, and the chopped fish settled into place with a squelch.

The boat's engine rumbled low and steady while Gossen continued to prepare for the morning's work. The gas tank was already filled, the oil already checked, the boat already given a good once-over before anything else got done. With everything taken care of, he glanced over the water sparkling in the morning sun as he prepared to climb into the boat. It was another great morning.

An investigator the rest of the year, Randy Gossen lives for crawfish season. It is his time. A time to be outside. A time to think. A time to watch the slow turn and change of the world. Two nearby television antennas that tower over him went up as he watched from the seat of his boat. The land from which they rise is not being farmed now, but somewhere in Gossen's eyes there was a long view of things that saw a tractor, or some other machine not yet imagined, one day turning the soil in order to coax rice or soybeans from the land. If that future machine materializes to do that, perhaps there will be a chance to coax crawfish out of the land too.

Randy Gossen himself does not farm. He works in collaboration with his cousin Dwayne Gossen. Dwayne farms more than a thousand acres each year. Some of it is family land; some of it belongs to others, who have placed in him their trust to make the best crop possible. Some fields he will plant with rice, and others soybeans. Some of those fields will rotate between those two crops in years to come, but others he will rotate between rice and crawfish, and in those fields he places his trust in his cousin.

Crawfish are not a crop like rice or soybeans, and they have largely, as we will see later, resisted easy understanding. Wresting them from the ground successfully comes from years of patient observation as well as individual trial and error. Anyone who crawfishes can tell you that, even with considerable hard-won know-how, getting the crawfish reliably out of a field and into a sack is nothing to be taken for granted.

Randy Gossen stooped under the boat's canopy and stepped in. He slid the tubs into place, wedged his tall frame into the driver's seat, and pulled the sorting table back on its rails so that it was within easy reach. With everything in place, he throttled up the engine and slid the boat backward off the land and into the water. With a practiced sense of timing, he flipped the lever that switched the boat into forward motion and began his first run of the day. Ahead of him lay a string of crawfish traps, spaced approximately forty feet apart.

With the boat itself setting a slow, deliberate pace, Gossen picked up an empty trap he had left near the beginning of his run and baited it. The trap is made out of nylon-coated steel mesh and looks like a four-sided pyramid, a tetrahedron, with a large, cylindrical chimney coming out of its top. At each bottom corner of the pyramid, the mesh has been pushed back in on itself, forming a funnel that opens into the body of the trap.

Properly placed, usually anchored with a steel rod but sometimes only carefully set down, a trap sits flat on flooded ground with the funnels offering an easy entrance to its interior. The bait is the welcome sign to the crawfish, who, having made their way in, cannot get back out. Their exit comes as Randy plucks a trap from the water, empties it from the top, rebaits it, and then places the trap not where it was but where the next trap is, as it itself is plucked from the water to be emptied, rebaited, and then replaced. The boat never stops. Its engine's roar changes rarely.

Gossen proceeded along his first line of traps, his body quickly remembering the rhythm and tempo of the work. The light breeze occasionally pushed at the boat, sliding sideways over the water, and he responded with a deft tap of his feet on the steering pedals beneath the sorting tray. At the end of the first line, a steady push of his left foot on its pedal turned the boat leftward, where more traps lay waiting. This morning, Randy began by working the line of traps at the perimeter of the cut—as the small, leveed-off sections of rice fields are known. As he approached his starting point, he turned in and started working the next line of traps in the forty-acre cut, following what amounted to a large, oblong spiral.

Trap upon trap, the work is steady. At each trap, Gossen leans a bit out of the boat and reaches down with his right hand to snare the rim of the trap. He switches the rim to his left hand as he picks the trap up and uses his right hand to dump its contents into the sorting table in front of him. He switches hands again and digs for a piece of slowly thawing, and increasingly stronger smelling, fish and drops it into the trap before placing the trap back into the water. That done, he has time to regard the contents of his catch, surveying the crawfish—Are they getting bigger? Have they molted recently? What price will this lot fetch?—to determine what changes he needs to make, if any, to his operation and to pluck out weeds and any other detritus that have come up in the trap. Today, the catch was reasonable and Gossen was enjoying the steady, if also a little slow, accumulation of crawfish on the table. Every few traps, he opened the doors to the chutes that guide the crawfish into the waiting sacks hanging off the table and then he cleaned the table of any remaining bits with a deft swirling motion of his hand that caught everything in it.

After about a half hour or so of steady work, Gossen had a sack of crawfish already tied up and lying on the bow deck of the boat, and he had two more sacks that were close to full hanging off the sorting table. This part of the field was done and it was time to move on to the next.

Gossen continued on in this way, adding thirty acres to forty acres to twenty-five acres, slowly working his way across the entire field. The work is always the same, but the views change as the boat moves around and the sun rises. With luck, the cool breeze and the warm sun combine to make for a pleasant day. On other days, the wind blows cold and hard and picks up an impressive bite as it crosses the water and slams into the boat's slab hull, pushing it about. Toward the end of the season, the breezes die away and there is only the heat growing heavier as the day wears on. And then there's the rain.

But today was a perfect day. Trap after trap. Line after line. Cut after cut. Each rhythm combined to make the time pass quickly until the moment came to move from one field to another, and that was when something amazing happened. Not so amazing for Randy Gossen, who does it many times a morning when crawfishing, but amazing for anyone else who might happen to be standing nearby and watching: Gossen pointed the boat at a corner of the field that led to the road where his truck sat. The boat dutifully took his directions and quickly ran its bow up onto the dirt at the field's edge, pushing water in front of it, slicking the dirt into mud. Most boats would have stopped there and awaited the pull of an arm or a winch to beach it thoroughly, but Gossen drove the boat further and further onto land, with only a slight pause to give the engine a bit more gas and to operate a hydraulic ram.

And then the boat heaved itself onto the land, exposing for the first time the wheels just behind its bow and demonstrating quite forcefully the power of its drive unit, which was not a propeller, but rather a large, cleated steel wheel that rolled the boat down the road, where Gossen turned and dropped into the next field.

If you asked anyone around here what kind of boat that was that you just passed on the highway—which can sometimes happen when fields are close—the answer would be a simple one: it was a crawfish boat. Crawfish boats were named after the job they make possible: they allow their operators to crawl into fields flooded with water and work at a human pace catching and bagging crawfish. Seen from a distance making their way through a field, they look like almost every other small fishing boat in Louisiana. Blunt bows give way to open, flat-bottomed hulls with a drive unit hanging off the back. If they watched them working their way through fields, a lot of people would hardly notice that the standard outboard is

Randy Gossen looking over his sorting table to make sure he is on target to pick up the next trap. Note the baited trap in his right hand ready to replace the one he will pick up.

Randy Gossen picking up a trap.

missing and that there is something else propelling the hull through the water. Most eyes do not linger long enough to see these boats slip out of the water and onto dry land, let alone ask how they got to be there.

And how they came to be there is an interesting question. The boats are not turned out in large quantities on assembly lines. They are not available at any boat dealer and cannot be found on any showroom floor. They cannot be found in any boat show. Anywhere. At any time.

Instead, they are made in small numbers in a handful of shops scattered across the south Louisiana landscape. And that is how it has been for the past thirty years. Some of the men who first brought the crawfish boat into being have retired from the work, but they were replaced by others who decided to try their hands at the task. Each has brought his own experience and his own particular way of looking at the world, which can be glimpsed in the way he conceives and assembles each boat he makes.

Their shops are littered about, sometimes on the edge of a town, sometimes along a well-traveled road, and sometimes at the end of a gravel lane that makes you wonder if you are in the right place. But in every instance, once you arrive, get out of your truck, and cross into the interiors of the large, often dark shops where men work on craft that stretch over twenty feet in length and sometimes eight feet in width, you find yourself welcomed with a ready handshake and a smile.

Those open hands and faces reflect a deeper reality: you are also face to face with individuals who have open minds and, not to get too sentimental, quite open hearts. These are men who are deeply in love with what they do, and are happiest when they are faced with a problem that only a piece of well-crafted metal can solve. And this speaks to one last point I want to make as we move forward into the history of the crawfish boat: the blue spark that you glimpse deep in these large metal buildings must really be understood as something created by the human mind, not simply the result of an unthinking hand pulling the trigger on a welding gun. The buildings that dot the landscape should not be dismissed as bastions of unthinking men bashing out bits of metal, but rather imagined as being akin to nurseries, places where the blue arc of creativity is protected, nurtured.

In reality, the arc of a weld is not so robust that it can withstand sustained winds. And it is dangerous when let loose in the rain. It leaves behind metal hot enough to do permanent damage should human flesh

Randy Gossen placing a baited trap ahead of the trap he is going to pick up.

Randy Gossen exiting one field/pond to drive down a field road into another.

come into contact with it, and injuries from working with so many sharp edges and powerful tools sometimes means that men work while dripping blood or biting their lips while a foot, finger, or head throbs from a slipped grip. Thanks to metal's thermal conductivity, the work is hotter than the heat of summer and colder than the frost of winter. And yet the work itself is so compelling that these men keep doing it. It makes them a living. It makes them happy. It makes them who they are. What follows is an attempt to understand the braided existence of minds, metal, and machines.

LAND'S END

This all started with the storms. In 2005, residents of Louisiana found themselves struck first by Hurricanes Katrina and Rita and then struck by the national debate that followed. The physical storms drove people from their homes; the discursive storms drove many more to despair. As residents watched for news of what was happening all around them, they were also privy to what others were saying about them, to the questions they were asking: Why did people stay? Why did they build there in the first place? Why would they want to rebuild there?

More than anything, the questions were about land. What was land and what wasn't land. It makes no sense, many argued, to build, or rebuild, a city, on land so, well, *not* land. The consensus seemed to be: too much water, too little land, too much risk. Even two years later, reporting on the state of things in New Orleans, *National Geographic* seemed to sum up the national consensus that emerged in the poststorm moment and had consolidated into a kind of truism with its lede to a story on the second anniversary of the storms:

> The *sinking city* faces *rising seas* and *stronger hurricanes*, protected only
> by *dwindling wetlands* and *flawed levees*. Yet people are trickling back to
> the place they call home, *rebuilding in harm's way*. (Bourne 2007, 33)

Those five adjective-noun pairs—*sinking city, rising seas, stronger hurricanes, dwindling wetlands, flawed levees*—build to a kind of apocalyptic inevitability that underlines the absurdity, and undermines the actuality, of living on what is clearly for outside observers a frightening landscape. Residents were, after all, *rebuilding in harm's way*. It was as if they were residents of Tokyo foolishly putting buildings back up after Godzilla's latest rampage.

More than anything else, it was the debate about the landscape, the very state of the land on which the state itself was built, that caught my attention. There seemed an enormous divide between the perception of nonresidents and the perception of residents. Where one group seemed terrified by the ambiguity of land that was both wet and dry, the other seemed either to accept the ambiguity or to gravitate toward the possibilities.

It reminded me of a rather bizarre psychological experiment I had read about that claimed that proof of human origins on the African savannah was to be found in the response of contemporary humans to certain environments. Testing hundreds of subjects by showing them images of different landscapes and asking them to give various responses to what they saw, the study argued that there are universal responses to environments. When shown pictures of meadows, test subjects reported feeling happy and talked about walking through the meadow. When shown pictures of swamps, however, participants felt anxious, depressed even. I remembered reading the study's report and digging in the notes to discover who had been polled. Sure enough, the subjects were all drawn from the same urban environment, in the Northeast, I believe, where the psychologist himself lived and worked. No wonder, I sighed. What do those people even know about marshes and swamps? My father may not know much about meadows, but put him in a boat in the middle of a swamp with enough cold drinks and snacks to get him through an afternoon, and he is as happy as can be, no matter whether any fish get caught.

Even from such anecdotal evidence, it seemed obvious to me that the landscapes that one inhabits have a significant role in shaping how you understand not only your own landscape but those of others as well. Having grown up in Louisiana and later living in other parts of the country, the differences in possible perspectives were not lost on me. I remembered all too well when I first moved to southern Indiana to go to school. One of the first things that greeted me, and delighted and baffled me, was the journeys one could take through the earth itself. Roads rolled up and down hillsides, but then, faced with too sharp an incline or too curved a path, they just cut right through the landscape, and I found myself driving through great arcs of limestone. Every semester, geology students would gather in these cuts, and while passersby traveled from one town to another, they would travel millions of years back in time, back to life's origins in the Paleozoic and Mesozoic eras.

Seeing those great rifts in the earth, I remembered overhearing conversations among neighbors about basements for buildings that could not be completed because they had struck bedrock. Growing up in Louisiana, I had never imagined that *bedrock* was anything more than a metaphor. It was as unreal to me as the doornails to which the dead are so often compared. Bedrock was a part of the landscape that could not be transformed, which for a boy from south Louisiana is both a mystery and something of an idyll. Having gotten stuck in my fair share of mud holes that looked more solid than they were and having come close to stepping into water so covered in duckweed or water hyacinth that it looked like an extension of the land that it bordered, I had always thought of land as fundamentally ambiguous.

It made me realize that, on a map, Louisiana looks like a capital L. The southward stroke starts broad, reaching from Shreveport through Monroe to Tallulah, narrowing as it sweeps past Alexandria. The eastward stroke seems almost equally wide as the one before it, but looks can deceive: our letter is not as solid as it appears. Nor is it as plain. Instead, it is elaborate, fringed with cheniers and ancient levees reaching out into the Gulf of Mexico. None of this is obvious to the casual eye surveying a map. But to those who live there, or who have traveled there, there is a vivid sense of how tightly Louisiana Highway 1 hugs Bayou Lafourche as it makes its way to Grand Isle. Sometimes the shoulder is as narrow as a teenager's on his first date. The same is true of any number of roadways reaching as far down as they can from the coastal prairie land into the gulf itself; highways and waterways wrestle each other for right of way, pride of path.

But as the *National Geographic* article made clear, reflecting a larger national discourse, such a landscape is alien, and alienating, to many. Too much water mixed up with the land.

The most common misconception of south Louisiana is that much of it is naturally occurring wetland. Even sensitive observers can miss the truth. Folklorist Alan Lomax spent his youth crisscrossing the country with his father, recording songs and melodies that they both feared would soon be lost forever in the face of the onslaught of commercial music. His exposure to an earlier American landscape made it possible for Lomax to return again and again to places he knew were rich in intelligence and beauty that other observers might overlook. His commitment and the wealth of materials he made available inspired many, including myself. He spent considerable time in Louisiana during those tours, and it was only natural that later, when he embarked upon his series of documentaries

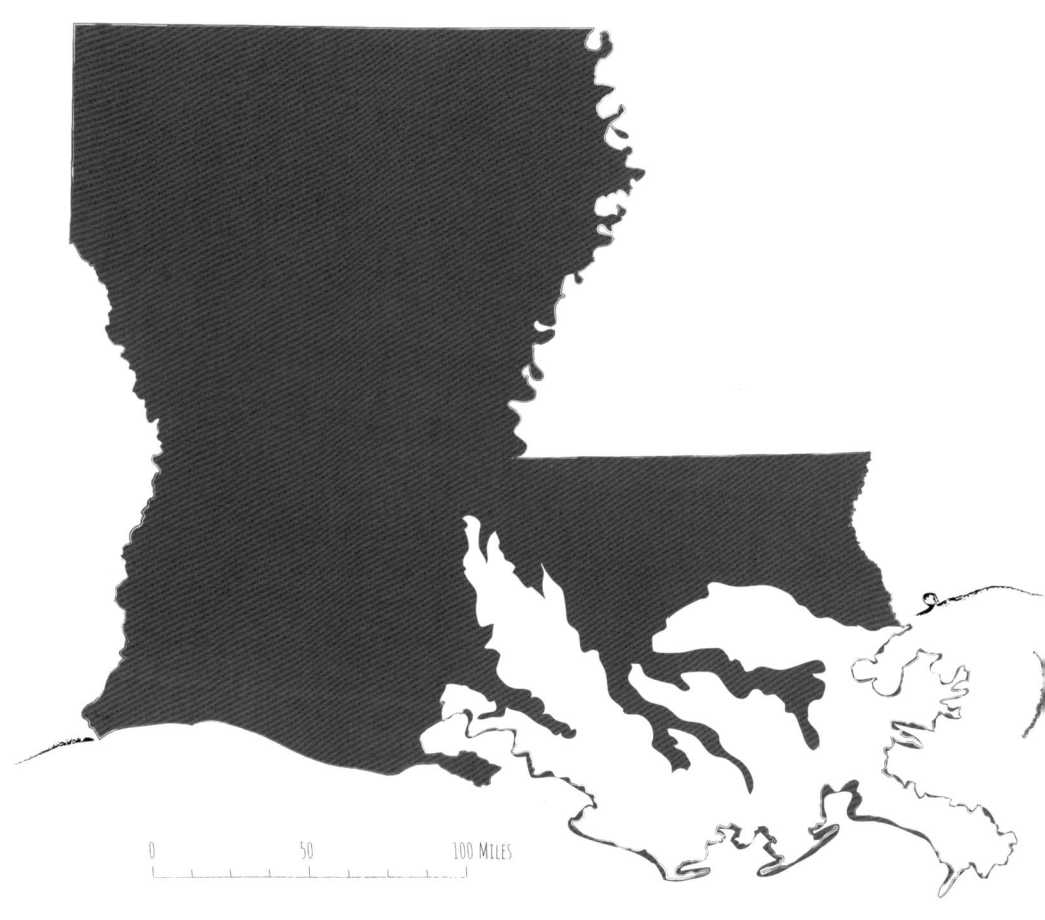

0   50   100 Miles

Map of Louisiana revealing those parts of the state where wetlands predominate and give way to the Gulf of Mexico.

celebrating what he called "America's patchwork culture," he would return to Louisiana, marking his time there and making it the subject of not one but two of his five films: one on jazz in New Orleans and one set in what is commonly known as "Cajun country."

The latter film is in fact titled *Cajun Country*, and it begins by first introducing you to a number of musicians, like Cajun fiddler Dennis McGee and Creole fiddler Canray Fontenot. Lomax gives a quick version of the usual history of the area and announces to his viewer that the film's subtitle is drawn from an old Cajun saying, "Ne lâche pas la patate" (Don't drop the potato), which is a local saying sometimes still heard to remind someone to hold onto the things that matter. As if seeking to ground his discussion in the landscape that provided the proverbial potato, Lomax plunges the viewer onto a fog-darkened road. Two signs creep past the camera. The first says "Mamou 18" and the next "Soileau." Mamou and Soileau are places, one gathers, somehow to be founded in this misty, mythical land. Strange-sounding places, but places nevertheless. You're in one, Soileau, and you're headed toward the other, Mamou. Lomax confides to the viewer: "I want to share with you one of my most extraordinary experiences: driving down a misty road, past shining silver marshes that are so typical of that area. Of course, it's all low-lying. You're always draining water so you can farm. It's a rice area."

There is no marsh between Soileau and Mamou. It's a rather high part of the larger prairie system, about seventy feet above sea level in Soileau, that, as you follow the highway, plunges down to a mere thirty feet above sea level when you cross the Bayou Nezpiqué, and then slowly climbs again to around sixty feet as you drive the remaining miles to Mamou. Such changes in elevation do not reveal a striking topography, but, rather, a more gentle one, one requiring some time to come to see its rounded ridges and soft swales. The road ahead in the film is indeed misty, and much of the landscape as one travels around south Louisiana looks like what is seen through the lens of the camera: roads curve and bend to follow bayous and rivers. At other times, they cut straight across a landscape that everywhere seems to be inundated with water. In the summer, especially, grassy rice stems crowd the water, making the fields look a lot like marshes to the untrained eye.

But it isn't marsh, it's prairie. Buffalo once roamed here, later replaced by herds of cattle when Cajuns ranched the area, and, now, a range of

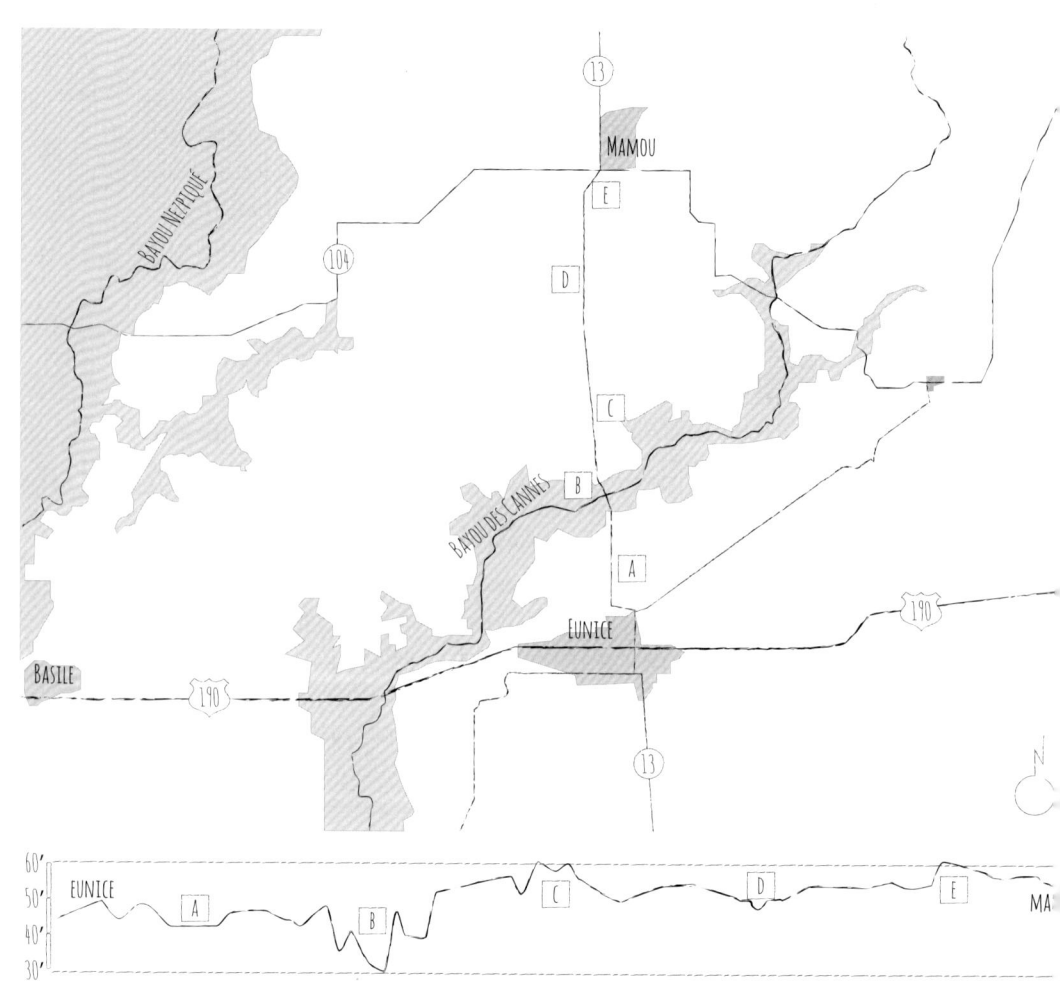

View of the route Alan Lomax took to get from Eunice to Mamou and the changes in elevation he experienced without necessarily realizing it.

descendants of Cajuns and Germans farm rice, soybeans, and crawfish. Many of the farmers insist that the land is not good for much more than that because the topsoil is so thin: in some places it is only a few inches deep. Beneath it is a tough claypan that is impermeable to almost everything: water holds on it or runs off, and trees have a hard time rooting in it.

Like many visitors to Louisiana, Lomax was not there to see the land. The fields he describes as marshes were mere backdrops for what he was really there to see: the "bedecked riders" of the *courir de Mardi Gras*, which is sometimes called the country Mardi Gras or the Cajun country Mardi Gras—although it is also practiced in some Creole communities as well. (There is a Creole one in Soileau, for example.) Any Louisiana Mardi Gras, no matter whether city or country, is spectacular. That is its job.

The spectacular nature can consume and confuse people, as folklorist Barry Jean Ancelet has observed.[10] They think that that is all there is to see. Were one to try to compile an image of the state from its many representations in both fictional genres like novels and films as well as nonfictional genres like journalistic accounts and documentaries, two things seem to be intertwined: Louisiana residents appear to live in a perpetual party, and if they don't live either along a street in New Orleans or a plantation along a bayou, then they live in the swamp. Think about it. When was the last time you saw a film set in Louisiana that didn't involve New Orleans, carnival, or alligators slipping ominously into dark, murky waters? In virtual Louisiana, there is no northern part of the state that joins with the rest of the south. There are no ordinary people living in suburban ranch homes with quaint, brick facades. Instead, everyone lives in a somewhat dilapidated—maybe *careworn* is the word that appears in scripts—wooden shack. They speak with an accent you haven't otherwise heard. If they're not drunk, then they are probably involved in some odd, religious practice that produces a similar altered mental state.

Many of these frequently occurring representations of Louisiana can be glimpsed in a film like *Southern Comfort,* which features a band of English-speaking national guardsmen fighting to survive a gauntlet of Cajun hunters who are angry that the guardsmen have stolen, of all things, their pirogues. The resemblance to the previous decade's *Deliverance* (1972) is pretty overwhelming, with toothless, sodomitic mountain men being replaced by French-speaking swamp dwellers who also know how to throw a good party, albeit one of the creepiest parties you'll ever see, with happy folk music playing as the two remaining heroes circle through a hunting camp, trying to kill their two remaining pursuers. The camp itself is, of course, located in the swamp, and as people stomp inside a tin metal shanty, our heroes wander among men boiling crawfish and slaughtering enough pigs to feed a small town.

The film repeats common notions that reveal more about outside anxieties regarding Louisiana and its people than anything about the place itself. Its 1981 release must have quickened the resolve of the two young scholars, Ancelet and the historian Carl Brasseaux, who had set out to try to undo some of the stereotypes. Joined by others in a research program that, under the auspices of the Center for Louisiana Studies, was ambitious in its depth and range, Ancelet and Brasseaux focused their efforts principally in giving the Cajuns and Cajun culture a proper history, with Ancelet focusing on collecting forms of expressive culture and Brasseaux seeking to establish the actual history of a group that had largely come to be imagined as Louisiana's alternative other: not black, but still similarly stigmatized.[11]

What the Cajun studies scholars wanted was an acknowledgment of the contribution of the communities who had, acting through the dynamic and durable fabric that only a collective of individuals can have, survived the crush of modernity and, in doing so, transformed Louisiana into a "folklore land."[12] Many of these communities existed then, and continue to exist in the present, at the fringes of the usual economies. Some of them seemed almost antimodern in their attitude, seeking to maintain ways of living that emphasized family and community over individual opportunity. This is not universally true for any of the groups, but it is somewhat ironic that the groups that were usually seen as most troublesome when it came to certain political and institutional notions of progress are now the ones for which the southern region of the state is most known.

Like other folk groups in other parts of the country and the world, Cajuns and Creoles are closely tied to the land in the imaginations of others—be they tourists or journalists or filmmakers. And so it was perhaps inevitable that identifiably strange groups such as these were tied to the identifiably strange parts of the Louisiana landscape. And so, despite historical research that has established that most Cajuns, something like 90 percent, lived, and continue to live, on fairly ordinary landscapes consist-

The graceful, curved lines of rice fields, especially of the internal levees that separate one cut from another, reveal the shape of the changing topography of the landscape that perhaps can only be known when you are trying to make things level.

ing of fields and rivers or, increasingly in the present, in tidy suburban homes in towns and cities of all sizes, we continue to pair the two.

Such a pairing, by naturalizing people to a landscape, relieves us of the burden of asking how people got there, what they do there, and what they think of the there in which they happen to be. Perhaps just as significant, by identifying a people with a landscape we also obscure who else happens to be on the landscape. And while it may come as no surprise that African Americans represent a significant portion of the population settled among Louisiana's bayous, prairies, and towns, it is also the case that there are a surprising number of German families, some of whose history in the region dates back only a little over a hundred years.

Many of the Germans, it has long been assumed, were slowly enculturated into the extant ethnic groups, much like what had happened along the Mississippi in the colonial era. But the Germans who had settled among the prairies and who had come in the aftermath of the Civil War arrived in a region undergoing, in some fashion or another, substantial changes in workforce dynamics, to say the least. And it was at this time that the Louisiana commercial rice industry emerged. It is clear that, thanks to both of the world wars, there was considerable pressure, both official and unofficial as well as internal and external, to Americanize, but much of that acculturation would have focused on matters of language. And I think this is where it gets interesting; like the Cajuns before them, observers have largely focused on the verbal and musical expressions of a group as the lens onto their worldview. If there have been material artifacts included in surveys of folk culture, they have been handmade objects such as quilts and pots and, occasionally, houses.

But what about the land itself? We use the word *agriculture* all the time, and yet those who study culture rarely think to, well, think about those agricultural practices as cultural practices, unless of course we are watching someone cut wheat by hand or work a field with a plow pulled by a horse or ox. Is it because, at least on the American landscape, agriculture is so far from us? Even in a city like Lafayette, where a half-hour drive from the city center can put you in the countryside among fields of sugarcane, rice, or soybeans, one rarely thinks about what is happening out there. To some degree, this limited view we have is a function of the powerful technology now available to farmers, such that one doesn't drive past fields filled with people planting or cultivating or harvesting. Instead, one occasionally glimpses a large tractor roaming around an otherwise empty field.

But does the difference between a field full of people or just one tractor make the process any less cultural? At the heart of that tractor is a human mind, and that mind is working the field in a way taught to him by his father and his neighbors, and his work will be evaluated by neighbors and by knowing passersby who will direct comments his way wherever farmers gather: at feed and seed stores during planting, at equipment dealers and repair shops during the working season, at mills during harvest, and at bank dinners, agricultural field days, and churches during slower times of the year.

No, the farmer is never alone. Whether he is riding high up in the cab of a tractor or circling yet another field in his pickup truck, he is constantly assessing the state of things, and possible vectors into the future, based on his own experiences as well as the received wisdom of others. Although a number of farmers with whom I spoke possessed university degrees or had attended college for a time, and although many regularly attended field days at the local research station and workshops hosted by various agencies, their day-to-day workings were based on the rhythms they themselves had developed from countless rides with a parent in a tractor on school afternoons, and later tedious hours spent doing things for reasons they didn't at the time understand. This is how farmwork enters the body. Slowly, accompanied by aches and bruises and the occasional cut or gash. The aches come from bouncing in the cab of a tractor or bouncing in the cab of a truck for hours at a time; the bruises and cuts come from wrestling a reluctant piece of gear into place and then having it suddenly snap into place—in such moments, bruises are welcome, since no flesh is left behind.

Intimacy is, perhaps, not the metaphor one expects to find when describing the relationship between working men, large machines, and the earth. But anyone who imagines anything otherwise is far wrong in their assumptions about the nature of this work as to be perplexing to the farmers themselves. Of course, they care for the land. It is the source of their livelihood. It pains them to see land poorly managed. Most farmers in this part of the world own only a minority of the land they work, and more than anything they prefer to rent land from owners who love it as they do. Sure, indifferent owners will leave them to their own, but it also means that they have done so in the past, which means that the land now in their care will need extra attention, if only because they don't want the land to wreck their machines, which can happen if a plow strikes a stray pipe or large piece of concrete.[13]

Most farmers don't mind the extra time they have to spend; it gives them a chance to get a feel, quite literally, for the lay of the land. Far from imagining the land as either wet or dry, they imagine it as a series of fields divided into cuts. Each cut has its own topography of "hills" and "holes" that will make flooding the field or holding rainwater a winning or losing proposition. They will spend hours driving around in a field, pulling a plow or a level, trying to get a sense of that field. Each bounce of their seat, each strain of the engine, puts them in immediate, intimate contact with the land beneath their feet in a way that most cannot imagine. A car striking a pothole while driving down a street is analogous, but few drivers will circle back to go over the pothole again and again, trying not only to determine its precise shape and size but also the best way to repair it. Instead, if the pothole enters the mind at all, it is as an object of irritation, an object that does something to us, to our car. It is not something we act upon, but something that acts upon us. Our response is not to consider its repair, but to defer its maintenance to someone else.

The relationship between farmers and land, then, is not only intimate and immediate but it is also active. The eye that roves over fields is part of the body that feels for hills and holes with the tools at hand. The tools themselves are not inert implements but powerful machines, tractors with horsepower ranging in the middle hundreds in many instances, pulling plows, levels, or cultivators thirty feet wide or more. And these devices themselves are made up of rotating disks or blades that move up and down not only in response to the earth itself but also in response to adjustments made by levers in the cab and transmitted to the equipment through hydraulic lines.

It is this active imagination that deserves discovery, an imagination that ignores passive distinctions like whether land is wet or dry for active ones like pumping water on or pumping water off. And it is in the fields where water has been pumped on, flooded up for rice and for crawfish, that we find boats that not only easily navigate a field that had once been dry but now is wet, but also leave a field behind, rolling up and out, onto a waiting road.

~~~~~~~

If, for the rest of us, there is some lingering concern about contamination, that land made wet cannot ever be trusted as land again, then the people living in south Louisiana do not share it; the crawfish boat is the nonpareil of an imagination that is not anxious about the transmutation of land and water. Wetlands are drained. Prairies are flooded, then drained, and then flooded again. A rolling landscape is terraced to hold rice and crawfish and low-lying fields are leveed to graze cattle.

Most important, an amphibious vehicle has arisen to allow farmers to become trappers, catching crustaceans that feast on last year's crop and selling them to an ever-expanding market. This machine fully participates not only in the natural landscape but also in the cultural landscape. There are, for example, no patents on any part of the crawfish boat. This is not because the men who make them are not fierce competitors, nor is it because they are unaware of intellectual-property laws or contemporary trends in patents and copyrights. In addition to his boat business, Kurt Venable mills a variety of custom parts for other manufacturers using his own CAM system. Gerard Olinger orders parts from his shop in the middle of Roberts Cove via his satellite service. Both of them are fully aware of the full force of the contemporary legal apparatus surrounding technology. On more than one occasion, Olinger has remarked that local fabricators always fill niches perceived as too small or unprofitable by large manufacturers. Both of these men, and any of the others, are fully capable of pursuing the legal steps necessary to mark some facet or another of the crawfish boat as belonging exclusively to him.

And yet no one does. As far as each maker is concerned, their reputations as builders, and the reputations of their boats—obviously, the two are intertwined—are well known throughout the community. Venable prides himself on making the strongest hulls, Richard on flexible hulls, Olinger on his dual-wheel drives. Each has also borrowed ideas from the others. Such borrowing is not always from direct observation but can often be in the form of indirect reporting: a farmer admires something on another farmer's boat and then requests that a maker add that to his own

~~~~~~~

boat. Sometimes the addition catches on more broadly; sometimes the logic of the addition or emendation is obvious to the maker in a way that leads to further innovation.

Creativity draws from this deep well of common knowledge and individual experience. Farming, like any other domain, presents a series of problems to be solved, but how those problems are solved are largely determined by how they are framed or understood, and that understanding is itself a function of individual and collective experiences that are constantly being negotiated not only in terms of content but also in terms of context. The framework for any solution, and thus the solution itself, is really a function of which individuals within a community are involved, which individuals have contributed, and who has accepted their contribution.

Each of the individuals in a community has to be understood as someone not only with particular abilities and self-perceived roles—only a farmer, a farmer who occasionally fabricates something when he needs it, a farmer who actively fabricates for himself and others, a fabricator who farms, or strictly a fabricator—but also in term of personal proclivities. For example, one fabricator is a tinker by personality, another is a born competitor and must win in whatever domain he enters, and yet another is raconteur of exceptional abilities. Together they make up not a homogeneous community, not even a cohesive one, but rather a loose network of individuals who, through their presence, maintain a network of ideas that have evolved over time. Those ideas are, of course, situated in a value matrix that has remained fairly stable for at least three decades, and it is reasonable to assume that this stability extends further back in time.

It is an ecology, and there could be no more striking example of the creativity of such an ecology than a modern metal machine gracefully wending its way through the water to the clatter of its small-bore engine and then lunging itself onto dry land, where it blithely rolls down the road to the next bit of water. This complex story of simultaneous invention and diffuse experimentation is itself set in a larger, unfolding social and economic matrix that is at the heart of modern American farming, where farm subsidies and price supports for crops are part of growing rice and but not of trapping crawfish.

The crawfish boat is an artifact born of modernity, but it realizes a number of traditional ideas within its various contexts. Traced through these various contexts, the artifact, be it a story or a boat, reveals it is always more than a thing. It always expresses something about the individual who made it and the individual who uses it. When those two individuals are part of a larger group with shared ideas, the artifact cannot help but express something of that culture as well as the landscape on which the group resides and the artifact operates. It is the peculiar charm of the crawfish boat that its destiny was to be born of an ambiguous landscape. Its mobility, no matter the circumstance, allows us a glimpse into how creativity has been practiced in a particular place at a particular moment in time. Perhaps no more, but certainly no less.

To understand the minds of these makers, we must first understand the landscape on which they work, followed by an examination of one particular set of mechanics, hydraulics, with which they approach problems. With the landscape and machines before us, we are prepared to witness the birth of the crawfish boat as it slowly develops into its current shape. That development is a function of men operating in loose networks of both discursive and material exchange that overlap and change shape during the course of the boat's thirty-year history. Our understanding will be based on a close examination of the thing itself, and from that, we may be able to glimpse how the minds of these men work. Of course, they understand themselves simply to be solving obvious problems posed to them by the constraints that we all face: the place in which they live and the time in which they live. We begin with their place.

SHOPS

From the edge of the field, I had a fine view of Randy Gossen's handling of the strange, amphibious craft he plied up, down, and around several hundred acres of Louisiana prairie. His handling of the boat was so graceful that it seemed less like he was driving a piece of machinery and more like a pas de deux of man and machine wherein the machine so fully embodied the ideas of its makers that it seemed a natural extension of its user.

It helps in this instance that makers and users are really not that far apart: the man who made this particular machine is Gerard Olinger, a more distant relation to Randy than his cousin and partner Dwayne, but a relation, and friend, nevertheless. Much the same can be said for the other men who make boats much like this one, both in the present moment and in the past. They are from this place and of this place, and one of the keys to understanding them surely must be to understand the place itself.

If you wanted to meet the man who made Randy Gossen's boat, you would need only to drive a mere three miles down a few blacktop roads to reach his shop. But if you wanted a crawfish boat, you have to go to where they are made, in a handful of shops, or the occasional equipment shed, in south Louisiana. None of the shops is large: almost all of them are one- or two-man operations, with most being family affairs of two brothers, a husband and wife, a father and son, or a brother and sister. Despite their size, or perhaps because of it, they are a vibrant part of this landscape, as critical to it as blacksmith shops once were.[14] As capable as smiths once were at wresting elaborate and useful shapes out of raw metal, the modern shops are called by a variety of names: Mike's Aluminum Welding, Venable Fabricators, Olinger Repair Service.

No matter what the specialization is behind the personal name, the shops use many of the same pieces of equipment and contain the same kinds of material. All of them have at least one welding rig—most have two, one for aluminum and one for steel. All of them have torches and saws for cutting metal to shape and a variety of tools for bending it or rolling it. The metal they cut, bend, and weld is pulled from stock pieces stowed on racks or on pallets or sometimes in what seems like untidy heaps on the floor inside or the ground outside. On those racks and in those heaps is metal of various hardnesses and thicknesses that comes as flat bars, angles, channels, tubes, pipes, or sheets. The aluminum can be plain or bright. The steel ranges in hardness and is, on occasion, stainless.

Out of this matrix of parts is built a variety of tools, devices, and machines: weirs to hold water at a certain level; pumps to move water into, or out of, a field or canal; ditchers to shape the land so that water will move where one wants it to; and boats to glide upon the water and scoop up a bounty that seems like a surplus delivered from the hands of fortune into the hands of farmers desperate for extra revenue when crop yields are off.

Despite such amazing fecundity, these shops have largely been ignored, perhaps because they live and work in the interstices of academic concerns. As organizations, they are too small for most business schools to examine. As gatherers of labor, their small size does not require much of a workforce, like those found in factories and large plants, and thus they are too small for sociology. As small businesses, they are only interesting to economists in aggregate: their individual revenues are small, as are the markets they serve. As producers of tools and machines, their creativity and inventiveness would seem to lie outside the arts and humanities, which too often seek intelligence in a category of objects they call art but others might simply consider useless.

Even in my own discipline of folklore studies, I suspect that the fact that they fuse metal with high-powered arc welders is too modern. Steel and aluminum cannot be gathered in the world, plucked from a landscape, and readily transformed into artifacts. Instead, steel and aluminum must be ordered from suppliers and delivered on large, flatbed trailers pulled by powerful diesel trucks that belch smoke as they pull up to shops that are themselves filled with smoke drifting up from fresh welds. Everywhere there is the tang of metal being worked, its edges either sharp or smoldering hot. Steel and aluminum are, by necessity, part of the larger global economy, which forces a kind of complexity to our work, an indebtedness to modernity that would seem to undermine our interest in the local, in the human. And yet what these shops do best is work on a human scale. For that is what these shops are: human. So fully human that most of us do not even notice them as we drive past their open bay doors, flinching at the bright-blue sparks of light and the loud crashes of tools and materials.

The Olinger Repair shop is a large metal building with approximately five thousand square feet under its roof. Inside are three large vehicle bays, a number of work spaces, a kitchen area, and an office.

Standing inside the Olinger Repair shop, it was easy to see not only the work involved but the imagination too. Jerry Leonards, a farmer, walked into the shop with a piece of steel in his hands. Wrenched from its original place and out of its original shape—violently so from the look of its twisted and torn end—it was hard to discern what it once was. There are some glimmers of color through the caked-on grease and dirt, but it was impossible to determine if the color was faded paint or simply rust.

Leonards paused long enough for his eyes to adjust from the bright light of Louisiana's semitropical sun to the darker interior of the shop. As his eyes moved around, he found himself in the company of a few other farmers: one had stopped by to check on the progress of a job he left for the shop and another was taking a break from plowing to linger near the open bay doors. Both were taking time out from a day otherwise spent alone in a tractor or truck cab to talk with neighbors and acquaintances. In a corner of the shop, a coffeemaker sat. The other two men saw Leonards and converged on him, and for the next few minutes they were drawn into a conversation about the current price of rice and what kind of rice they planned to plant this year, in what fields, using which method.

Leonards is a familiar presence in the Olinger shop. He is a relative by some mathematics of genealogy, which almost everyone in south Louisiana can do and which the residents of Roberts Cove, with their dense patterns of overlapping families, can calculate in mere seconds. Leonards farms land near the shop, making it a convenient place for him to frequent. His glasses somewhat obscure his wide-open eyes, which go well with his curiosity and his willingness to laugh at himself as well as the world: he likes to make fun of himself as a farmer, once remarking, in the midst of a really poor soybean season that had everyone worried, that he was just going to plow everything under and be done with it. (Farmers' jokes have their own logic.)

It would not be unusual for Leonards to come into the shop and have no one approach him, the place seemingly deserted if it weren't for the burr of a grinder in use or the crackle and hum of a welder. Stepping carefully past large pieces of equipment, Leonards knew to seek out a pair of feet poking from under a tractor or grain cart or the glint of a welding mask behind a fiery-blue arc. This time, standing with a torn piece of steel dangling from his hand, Leonards wanted to order something, but he didn't have a set of drawings—very few of Olinger's customers do. What he had was an idea

Gerard Olinger milling an axle for a side plow.

in his head, sometimes only the roughest of ideas. He delivered it using the words he knows, which may not have included any proper terms, and a few simple gestures to conjure up the spatial relationships that could not be conveyed readily with words. Gerard Olinger—the man behind the mask—now raised himself up, listened, and watched the hands. He asked a few questions and then nodded his head.

His questions included not only those that covered the function and composition of the steel object but also about the urgency of the project and when was the absolute latest it could be ready. That is the usual limit of specifications for any particular project. There are no drawings for much of the work that happens in these small fabrication shops, only a few notes, often made on the back of a rusted piece of metal or sketched on the concrete of the shop floor. The only way their design work is captured and preserved is in the work itself.

And the work is judged on its own merits: Does it work? Does it hold up? Does it look good? There is talk, but the work itself is the conversation. The talk seems almost ephemeral, orbiting about the work, informing it but also informed by it. People notice the careful bead of a weld, which reveals that the tacit knowledge of an occupation lies not only in those who have mastered it but also in lesser practitioners and in the wider audience of users and morning-coffee evaluators.[15]

And so such fabrication takes place within a larger network of ideas and practices with which all the individuals are familiar but to which they have different kinds of access and with which they have different experiences. Any particular individual could be strictly a farmer, a farmer who fabricates on occasion and only for himself, a farmer who actively fabricates for himself and others, a fabricator who farms or is a member of a farming family, or strictly a fabricator. Out of this assortment of abilities and interests flows a steady stream of innovations and adaptations in response to particular problems: side plows, PTO (power take-off) ditchers and other powered machines attached to tractors, fixed and mobile pumps, small boat hulls (for airboats and hunting boats), surface-drive engines, and crawfish boats.

Fabricating, or making things, is a fundamental part of life in south Louisiana—as it is in many other parts of the nation (and the world). Looking at such making as part of a larger ecology reveals the complex web of relationships and full range of roles available to individual par-

A PTO (power take-off) ditcher in back of the Olinger shop in for repair. In most cases, a chain or a bearing needs to be replaced, but like a lot of equipment, ditchers can get used hard and require significant overhauls.

ticipants.[16] There are a variety of ways of approaching the task, but the way folklore studies does it is to go where people are most involved in the production, distribution, and consumption of the things they make.[17] This usually involves finding either smaller parts of larger systems or finding larger systems.

Proximity of ideas and their realization seemed obvious as Jerry Leonards left the shop with the part he needed in his hand. Leonards's urgent job taken care of, Gerard Olinger returned to his work assembling a PTO ditcher. A PTO ditcher sits immediately behind a tractor's cab, raised and lowered into place by the tractor's three-point hitch mechanism and powered by the power take-off spindle that is geared directly to the engine. They are amazing machines, spinning a cutting head of thick steel blades at several thousand revolutions per minute to throw any dirt they encounter high into air and far from where they are making a ditch through a field. In this way they leave no furrow to keep water from draining easily from a field, making it possible for farmers to work the water level in their fields with an impressive amount of precision.

PTO ditchers are but one of a half-dozen specialty items for which the Olinger shop is known. The Olingers do not worry about competitors, neither near nor far. For them, certain kinds of competition are inevitable, and should a large company seek to muscle them aside in a given market, the Olingers are confident they would find other things that need doing. The fact is, farms are complex operations with large, heavy, and complex machines that are shoved into extremely dirty and often dusty environments. They will break. The Olingers find their confidence not only in their ability to fix things, but also in their being a part of the landscape and respecting it and the farmers who work it. They themselves once farmed, only quitting to focus more energy on the needs of their neighbors, who come to them not only with things in need of fixing but also things in need of making.

And so the Olingers build a new tool or machine and another farmer sees it, either in their shop while it is under construction or in another farmer's field, and he stops to ask about it. He wants one, too. The Olingers make one, and then another. Soon they have made a dozen, and already some of the first few have come back—perhaps broken, or perhaps with a request for a modification based on a particular farmer's needs or his insights. Sometimes the Olingers themselves make modifications or improvements based on their own need to simplify a building process or to take advantage of a potential strength or efficiency they have suddenly glimpsed.

All this happens in situ, in this place. In their place, which is not simply theirs, as the steady stream of farmers into and out of the shop reminded me. There was, I thought, something of what the French call *terroir*, which typically means something like the flavor of a place, the idea that you can taste the character of the climate, the environment, and the earth in the product. Food critics use it to describe wines and certain kinds of foods, but there is something more to it. Dell Hymes once tried to define it in relationship to the American Northwest, as he puzzled over the similarities between Native American oral discourse and the poetry of Robinson Jeffers.[18] It is, if nothing else, a question to be asked rather than a fact to be assumed—if only because the original conception of ethnicity so clearly tied people to the land, and in so doing classed them out of modernity.[19] The best place to ask that question is when the people involved are working the land itself. In south Louisiana, that means creating terraced fields for rice.

WORKING THE LAND

I wanted to see the land for myself, and I wanted to see how the farmers worked the land. So one day I drove out to see someone water leveling, which is when a farmer floods a field before grading it as smooth and level as he can. Water leveling was once the only way the work was done, taking advantage of the ability to stir up a lot of dirt into the water so that it would naturally settle into a smooth surface, but with more powerful, and more precise, equipment, some farmers prefer to dry level, though it's been my experience that dry leveling tends to be performed when major changes are at hand: cuts within a field are getting enlarged or the topsoil is being swept aside in order to work the claypan.

On this particular day, I pulled off the road near where a large tractor circled inside a muddy pond. Its four pairs of wheels, slicked with mud, glimmered in the gentle afternoon light of late winter. Around and around the tractor went, as if it were trapped in the pond, a restless beast pacing in its confines.

Suddenly, as if sensing my presence, the great machine broke its stride and lumbered to a spot in the pond near where I stood. Up at the top, a cabin door swung open and from its interior a head popped out. Dwayne Gossen's gentle smile spread across his face as he pushed his sunglasses up on his head and waved me over. Gossen is a first cousin to Randy Gossen and a second cousin to the Olingers. He is also related, through an intricate genealogy that I do not quite follow, to Kurt Venable. While he is close to the Olingers, both because he farms family land and because his own farm is so close to their shop, he once worked for Venable, helping to build Cajun microwaves while in high school.

Dwayne Gossen is quite typical of many of the farmers of the area: quiet to the point of seeming shy, but a fount of warmth and hospitality when asked an honest question. (So many of these men are like this that one wonders whether the world would not be better if everyone took a turn at farming.) I leaped across the ditch separating the road from the edge of the field and almost lost my footing as my boot squelched and sank into the earth softened by Gossen's activity. A fortunate tilt forward brought my hand close enough to hold on to the tractor to steady myself and land my other boot on the lowest step. At the top of the ladder, Gossen motioned me to the jump seat that almost all manufacturers install in large tractors.

"It's nice to have company," he said.

Big tractors like this one, a Case Steiger 385, are not designed for small jobs. They are not a utility tractor that gads about a farm, performing any number of small jobs, the smallness being either a matter of the power required or, in the case of larger jobs, for the amount of time required. The tractor's number reflects its horsepower, and its eight wheels give it the ability to use its power to pull large implements, like the thirty-foot blade currently attached to it, all day long. Rarely is it the case that such days occur singly: when an articulated tractor comes out of its shed or off its spot in the equipment yard, a farmer is already committed to days, if not weeks, of long days that begin at first light and often end well past dark. The big tractors make the work, when glimpsed from the side of the road, appear effortless, but no one who has spent any time bouncing along with a farmer working a field, constantly turning both the tractor and his own body to check how things are going behind him, would call the work anything less than exhausting, an exhaustion usually made unavoidable by the narrow windows of time within which the men must work, ahead of or just behind whatever schedule Mother Nature is on for a particular year.

As I shut the door, Gossen revved the tractor's engine back up to its working speed and slowly allowed it to pull itself through the muddy water. Perched so high up the landscape on which he worked, it was easier to see that the tractor worked inside one part, called a cut, of a larger field, and that cut was separated from other parts—cuts—by a small levee. The field itself was enclosed by a larger levee, beyond which lay ditches and roads. All of the cuts were full of water, pumped from a well over the past few days to make the land more workable, following a tradition reaching back several decades of leveling a cut while it was flooded with water. It is yet another instance of an intelligent use of an abundant resource, water, to make the land more useful to humans.

Looking out across the series of flooded cuts making up several fields, I remembered a recent conversation with geologist Gary Kinsland, who

Dwayne Gossen leveling a field in late winter/early spring. This particular plow was fabricated by the Olingers and has wings that reduce its operating width of thirty feet to a width of only eighteen feet or so for ease of transportation between fields. Note the water ahead of the tractor and smooth mud behind.

started off by emphasizing that the way Louisiana was built could be a little confusing to most. We are used to land being created by events like continents colliding and volcanos exploding, but Louisiana, especially south Louisiana, is the product of a slow, even gentle, process of building layer upon layer of sand and clay and rocks, a variety of materials that reveal themselves to us in the gentle topography of the lower part of the state, where the highest point in this part of the state is 190 feet, and that's from the vantage point of a salt dome.

As Kinsland told it, Louisiana was built from the north. Much of what we think of as the state is, in geological time, quite young. When the two American continents separated, the sea we now know as the Gulf of Mexico was a much larger ocean whose northern coastline ran through the middle of present-day Arkansas. Eastern Texas, all of Louisiana and Florida, and most of Alabama, Mississippi, and Georgia did not yet exist as dry land.

Later, in the Paleozoic era, the two continents swung together again, as part of the larger supercontinent Pangaea, and the ocean, compressed into a much smaller space, became the basis for the state's current wealth of salt deposits. As Pangaea broke apart and the two continents were again divided, Louisiana was under water. As a shallow sea, it accepted the mud deposits, rich in carbonates, that would become the source of petroleum. As sea levels rose and fell, northern Louisiana began to fill with sediment, but the prairies of south Louisiana that run gently to the Gulf of Mexico were in deep water.

All of this changed with the increased volcanic activity and the meteorite that struck off the Yucatan Peninsula that ended the life of most of the dinosaurs. The Cretaceous-Tertiary extinction event, as it is known, marked the end of the Mesozoic era and the beginning of the era in which we now live, the Cenozoic. Over the next sixty million years, the southern edge of North America crept slowly southward, thanks to the steady flow of sediments through a developing network of rivers. The thing to remember, Kinsland assured me, is that although we mostly think of sediment as being the fine silt we see in contemporary rivers, it was not always so. In the past, a wide range of materials has come tumbling down riverbeds and then spilled across a seafloor. Those seafloors gradually rose as they were built up and ever so gradually fell again as the material of which they were made compacted.

Another reason for the falling, and which is an essential part of Louisiana's geological story, lies in the earth's crust, which gave slightly under the weight of the material piling on top of it, rather like, as Kinsland pointed out, a piece of Jell-O giving under the weight of your thumb. The combined effect is to create a series of nested bowls, each edge a little lower than the last. The lips of the bowls are terrace-like in appearance, and the whole stack tilts ever so gently toward the gulf.

All of this occurred as the Mississippi River slowly emerged as the dominant river in a complex system of rivers. Much of the materials that form the many layers of Louisiana are, depending upon the time period, from much further north. All the little round rocks found in Louisiana, for example, are called chert and are silicate rocks created three hundred million years ago that were eventually washed down from Canada through this complex and powerful river system. As the Mississippi slowly developed its course, it established, like many rivers do, its own system of banks, which were built up from the same materials it had previously, and so generously, distributed across the landscape. The great western levee of the Mississippi can still be easily discerned, albeit on the western side of the Atchafalaya Basin, about forty miles from the river's present course. It is known locally as the *coteau,* or sometimes as the slightly redundant *coteau ridge,* and is part of the names for several towns in the area, including Grand Coteau and Coteau Holmes.

"This all was built in the past couple million years," Kinsland reminded me. "A short time for geologists."

Seventeen thousand years ago, sea levels were three hundred feet lower. The Mississippi River valley between Lafayette and Baton Rouge once descended three hundred feet, so that the river could get to the gulf. When sea levels rose, the river almost filled in the valley, leaving the coteau. The prairies are between 85,000 and 125,000 years old and were created during a previous period of high sea levels. The lowest levels were during the last great ice age, but there were fluctuations between these two dates. The deposition during this period is similar to that going on now in the marsh areas.

Kinsland pointed out that the northernmost steps, sometimes also called *terraces,* are the oldest. At the bottom of the state, the newest area, are the coastal marshes. In between is the prairie terrace, or simply the prairie, which was built up during a period when very fine-grained sediment poured down the Mississippi.[20] Its surface is dense clay, whose low permeability makes the land hold water easily, an important element in rice farming, or, conversely, a fact of life farmers have accommodated by growing a crop such as rice that tolerates water.

Further north, Kinsland noted, is fluvial sediment. Its fine grains are buried, but what remains is sandy. Five miles north of Ville Platte, you arrive in a different world. The surface is permeable. The crops change. You are out of the prairies, out of the grasslands, out of Cajun country, out of the area that the geographer Fred Kniffen once dubbed "the French triangle."[21] This triangle of land, stretching from where the Red and Mississippi Rivers meet, is bounded by the Atchafalaya Basin to the east and by a southwesterly line to the west that results from the deposited material compacting, even to this day, as it slopes slowly to the Gulf of Mexico.[22]

As the last great ice age ended and the topographies and bathymetries of the current landscape were revealed, the prairies would forever be limited in their capacity to support life by the thick, tough claypan that underlay them. In places, the topsoil is as thin as four inches; in others, it increases to a few feet. In any case, there is not much there. Before European colonization, the prairies supported a few native grasses that fed buffalo and deer, which were in turn preyed upon by various Native American groups. When the Cajuns arrived on the eastern edge of the prairies, they mostly ranched the region, with subsistence farming sprinkled throughout. Only along the rich banks of the bayous could the plantation system thrive. Further west, the land simply couldn't provide much of a base for sustained agriculture.

That is, not until folks like Dwayne Gossen's ancestors arrived and began, purposefully, to flood the land to grow rice. They built levees to hold the water in, and they divided the fields into smaller cuts in order to make each area as level as possible. The gentle topography of Louisiana makes it harder to see these "rice terraces," which are similar to the ones seen along mountainsides in other parts of the world, but they are terraces nevertheless. Like terraces elsewhere, their purpose is to control the distribution of water.

The idea behind breaking up any slope into a series of steps is to make each step as level as possible. The importance of this becomes clear the first time you see a rice field drained: a properly leveled field will be clear of water across its entire breadth. Uneven patches will result in some areas of a cut holding water. Conversely, when a farmer floods a field early in the growing season and only wants to hold an inch of water over his newly sprouted rice, he does not want to see small hillocks of dry land where plants will not get the benefit of his careful and, depending upon energy costs, sometimes costly pumping. Someone as careful as Dwayne Gossen does not want to see more than an inch of variation across an entire cut, and getting forty acres of land that level requires a great deal of work.

This work consists of dragging a large grading blade, essentially a much wider version of a bulldozer blade, behind a tractor. A few farmers do this when the fields are dry, known as *dry leveling,* but the traditional way is with water pumped into the cuts, appropriately called *water leveling.* Before the advent of laser leveling systems, water leveling worked through a combination of a farmer's skill at handling the blade and moving the dirt around, his knowledge of the land, and the simple physics of suspending a fair amount of dirt in water, resulting in an even dispersal of the mud across an area as it slowly precipitated out.

Gossen finished the field he was working in and pulled to the side of the field, not too far from where he had parked his truck. We climbed out of the tractor and into the truck. We pulled into the driveway of the nice home next to the field and drove down to where a trailer, its elevated platform, and an adjustable pole sat. Gossen got down from the truck, opened the back door of the crew cab, and fetched out a yellow, hard-plastic case. He walked to the trailer and climbed up the steps to the platform. He unbolted the beacon, a fairly expensive piece of equipment, from a post braced in the middle of the trailer. He opened the case and removed a laser transmitter, which was about the size of those old-fashioned flashing lights on police cars from earlier in the twentieth century. The transmitter works on a similar principle: a laser mounted inside the body of the transmitter shines into a rotating reflector, creating an invisible, at least to the naked eye, plane of light a set number of feet off the ground. We then cranked up the four feet at each of the trailer's corners that not only stabilized the trailer but also made it possible to get it closer to level than when it is simply parked somewhere. Rarely do farmers have the luxury of being able to put a trailer on level ground that also happens to be conveniently placed near where they are working, so the beacon's mount gives some flexibility, but considering that trailers are often wedged between a

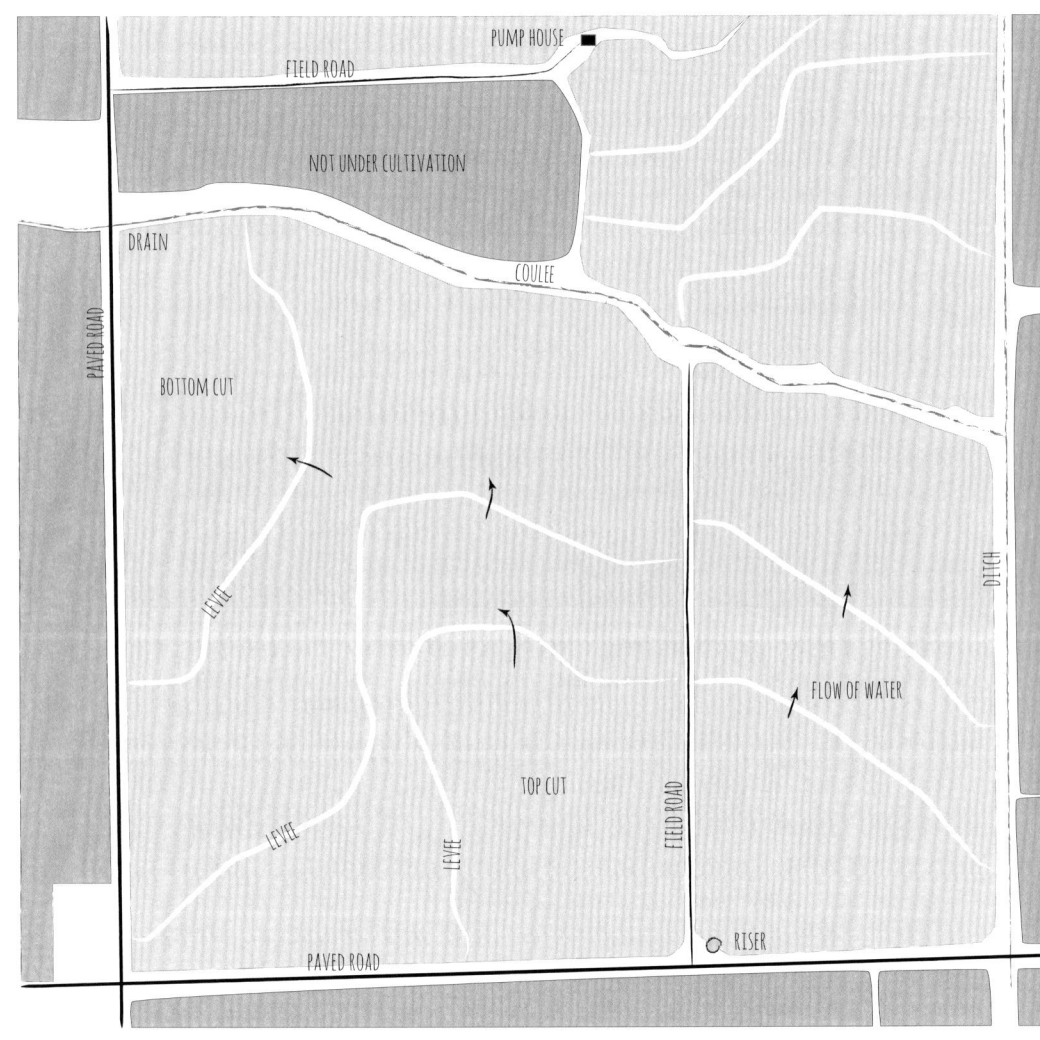

Map of a rice field showing the flow of water from risers at the top flooding upper cuts to lower cuts draining into ditches or coulees. Note how the complex curves of the land's topography is fitted within the straight lines of roads and property boundaries.

roadway and a field edge, the chances of starting from less than a fifteen-degree tilt are not great.

After hooking up the trailer, we got back into the truck and headed down the driveway. A woman walked out of the house and Gossen stopped and rolled down his window.

"Thanks again for letting me park my trailer. I hope I didn't mess you up in any way," he said.

"Oh, it's no problem," she said with a nice smile. It was clear she enjoyed his good manners. "It's all God's country and you take such good care of it."

After the exchange of these pleasantries, we were off. Gossen told me he always tried to do the right thing, to thank people when they were helpful and to try to limit the mess he made as he worked, especially when people's homes were so close to the fields he worked.

We drove west down the highway and turned left onto a narrow, black-topped road. Gossen found the next spot where he wanted to put the transmitter trailer and deftly backed it right about where he wanted it. He reversed his steps, grabbing the yellow box out of the backseat and stepping onto the trailer with its four-foot-high telescoping pole. He bolted the transmitter in place atop the adjustable pole that itself was on a platform mounted about four feet off the trailer, leveled it using the built-in bubble, and cranked it up to what experience told him would be about the right height.[23] We got back into the truck and drove to where the tractor still sat in a field. Gossen asked me to drive the truck to a spot where it would be conveniently located when he finished the next field: his directions required sometimes translating locally known landmarks into something I could understand, often by counting driveways or some other obvious physical distinction, in his head in order to give me a number to go by.

With everything in place—the tractor and water leveler moved from one field to another, the laser transmitting with no visual obstacles to its swirling signal, and the truck awaiting us at the end of a series of cuts to be graded—Gossen once again got back to work. His first maneuver was to make a circuit of the cut, establishing high and low points in reference to the laser beacon. He did this by watching a series of numbers appear on a readout mounted on the right-front pillar of the tractor's cab. The readout's data flows to it via a cable that runs from the tractor to another post mounted on the top of the water leveler itself. A receiver, which amounts to a collection of eyes, moves up and down on the post until it locks onto the plane of the transmitting laser. With the lock in place, Gossen can choose to let the grader blade rise and fall with the wheels, moving the receiver up or down if it falls too far below or rises too far above the transmission plane, or he can slave the height of the grader blade to the receiver, forcing the blade to remain at a constant height while the rest of the water leveler raises and lowers to make that happen. Raising and lowering the instrument occurs via hydraulic pistons that are themselves powered by the large tractor, and thus the readout inside the cab is actually a fairly powerful computer constantly monitoring the underwater topography of the cut and making adjustments accordingly.

It all seems fairly automatic, and almost all farmers now rely upon the assistance of a laser leveling system. Before its arrival, a lot depended upon the eye, and feel, of a farmer as he drove around and around in a cut filled with a foot of muddy water. The laser leveler takes some of the eye out, but it cannot take away the importance of a farmer being able to feel what is going on with his topsoil as he swirls it around with a thirty-foot-wide blade hauled behind his tractor. It also requires a farmer to understand how to move earth around in order to maximize his efforts and minimize his time. There are, after all, more cuts awaiting his attention.

Up in the cab of the giant Case, the work seemed effortless as far as the tractor was concerned. Its diesel engine rarely seemed to rise above a fast idle, but Gossen was alert to changes in the roar and in the rumble. His head was in constant motion, checking gauges, checking the leveler's readout, and then twisting backward, usually with the upper half of his body following, to see how much mud he was pulling in the blade and how well the tire ruts were filling as he worked the soil.

"I like to make a few rounds before I start pulling," he said.

Managing the agricultural landscape is all about pulling. If you are rotating soybeans and rice in a field, then the first thing you do after you harvest beans is to pull up a levee in the autumn so that it will be hardened by spring. Conversely, if you are done with rice in a field, you pull down those levees. Inside a rice field cut, you are either pulling down hills or you are pulling up holes. To add to the complexity, a farmer needs to be careful not to over-pull or under-pull as he grades a cut. Much of this particular turn of language resides in the fact that all of this work is done with implements—plows, cultivators, levelers—that are pulled behind a tractor, and so in the same way we extend our own sense of self to include the instruments we use, as when we say *I sawed a board* and not *I used a saw to cut a board,* farmers in the region say they spent the day pulling—pulling things up and pulling things down.

The man and the machine are part of a larger whole, and good equipment operators are acutely aware of their machines, in almost the same way that a runner is aware of his shoe as an extension of his foot. Gossen can see neither high nor low spots in the muddy water of the cut. Instead, he senses things through the tractor itself, through subtle changes in pitch or vibration that alert him to slight changes in the amount of work the tractor is doing. He is also always twisting in his seat to see how much mud, if any, is creeping up on the blade and to see what the mud looks like behind the blade: a thirty-foot-wide piece of dished steel that is also twice as high as the water in a cut moves a considerable amount of water out of the way. In its wake it is almost possible to see the ground itself, without any water on it, before the water comes rushing back in from both sides.

After Gossen made his rounds to feel out the cut, he set the blade to lock its height to the receiver and made his first pass, tracing the perimeter of the cut. A series of slow and careful sweeps diagonally across the cut followed the edging, with each sweep always slowing a little toward its end so as to minimize the chance of water topping the cut's levee. How he sweeps, and at what angle he sweeps, is a function of the shape of the cut itself. Only a few cuts in any given field are reasonably rectangular, and those are usually at the very top or bottom of a field. All the rest of the cuts tend to wrap or wiggle around, following the nap of the land in an attempt to make leveling within as easy as possible. The cuts are bigger these days, and their shapes are simpler, a function of the much more powerful equipment available to farmers, but the land itself, in combination with gravity, still dictates what is possible and what is not.

The importance of gravity is in the role it plays in moving water through any series of cuts and fields. At the heart, then, of any collection of rice fields is a well. A pump draws water out of the well and feeds it into several large pipes, each usually a foot or more in diameter, that spider out to local high points. At each high point, at the top of a field or a couple of fields, is a riser, a giant spigot out of which gushes cold groundwater that pours into the top cut of a field. Each cut has at least one drain in it—the drains can be pipes or plastic curtains across low spots in a levee—that allows a farmer to distribute the water from the top cut, through a series of inter-vening cuts, all the way to the lowermost cut of a field. Each drain, be it a pipe or a curtain, is adjusted to just the right height for the particulars of the growing season, and the goal is to fill all the cuts with the right amount of water and not a single drop more—overfilling a field, especially after any kind of application such as fertilizer, is literally throwing money down the drain. Farmers not only care about their environs; they also care about their wallets. A heavy rain at the wrong moment results in some very col-orful conversations in equipment sheds throughout the area.

Gossen had been running his pump for several days to get the water up to a height that he could level his fields. He was particularly worried about the cut we were in. In previous years he had a hard time with a low spot in one of its corners. He focused his efforts on trying to erase it without overworking the cut.

Not completely satisfied with how things were feeling as he ran over and around the area, he said, "I don't know. I don't like to leave it like that, but there's only so much you can do at a time." With a mental note made to check the corner in two years' time, the next time he would plant rice there, he pointed the tractor at another corner of the cut and pressed the button to raise the water leveler. The machine moved easily ahead, seem-ingly enjoying some relaxation after an hour's tension of constant pulling.

Gossen continued to work cut after cut, finishing out the one field be-fore calling it a day. Each cut not only had its own shape, but also its own personality, if you will, which was a function of Gossen's experience of the cut over the years: what it wanted to do, how it tended to go. This person-ality of a cut or a field is something farmers carry around with them; they are discussed when they see each other at a local shop or after church or at a local community event such as fund-raisers or ball games. In some ways, fields have their own lineages that are not just a matter of who owns the land but who has worked the land as well. Sometimes an older farmer will know something about a piece of land being worked by a younger farmer because he was friends with the man who worked it before, and he can remember that a particular cut never produced much rice or that another cut always produced the biggest crawfish or yet another cut was never dry if the rainfall was anything more than average. The men carry these stories around and swap them, intermingled with stories about their children.

RICE FIELD CUTS & DRAINS

CURTAIN

SIMPLE PIPE

BURIED SELF-LEVELING PIPE

This cutaway of a rice field glimpsed in profile reveals how the levees and drains work together to create a tiered set of ponds in which both rice and crawfish can be grown.

Dwayne Gossen leveling a rice field. In this image, Gossen is in the middle of a turn and you can see the amount of water that these plows gather in front. Farmers have to slow as they approach the edges of cuts to ensure they don't slosh water out of a field.

There are a variety of drain systems used by farmers to move water from one cut to another. Some use curtains that allow water to flow over a levee, and others use pipes that allow water to flow under or through a levee. Here, a flexible pipe has been placed. Water is controlled by either placing a piece of square aluminum, seen here stuck in the ground to the left of the pipe, in front of the mouth of the pipe or by removing it. It's not as elegant as an embedded drain, but it is significantly less expensive, and it is easily removed for soybean cultivation.

The land can be as mysterious as a family member—and as recalcitrant about responding in a way you expect as another person.

This is the difference between land and many other things within the realm of nature, like the weather, which is perhaps imagined more as a force. Land is broken into discrete entities, like people into bodies, and thus it can be imagined that way, perhaps even managed if one recognizes that it has an existing personality that one may or may not be able to change for the better—and there is also a sense that it can be antagonized, pushed too far. The weather, on the other hand, is separated by years or, more accurately, by growing seasons, but it is too large to be cognizable in the same way as the land. Although I have stood under the shed roofs of many an equipment barn and watched isolated thunderstorms pour rain over one man's land and bypass another's entirely, farmers never imagine storms as discrete. Fronts may come, but to consider them distinct would make as much sense as thinking the sea is distinct because it crashes onto the shore in waves. For farmers, the weather is not a text but a context. What that means for them is that it is part of the larger environment within which you work. You focus your efforts where you can: in preparing the land for cultivation. You can only hope that the weather will be kind, the yields good, and the market strong. That is all a farmer can hope for. No more.

CULTIVATED

As she prepared to enter her school, my daughter clutched a rice plant I had pulled from a farmer's field the afternoon before (with permission, of course). Her fifth-grade class was discussing local geography and one of her words had been *terrace,* so I had promised to bring her back something from a terrace, honoring the days she has spent with me, driving around the Louisiana countryside. We have, she and I, been known to stop to walk on newly seeded rice fields, where she has traced with her small fingers the cracks of the earth formed as fields dry. She has walked many a rice terrace, teetered along the tops of levees, and pushed her hands deep into the cool water that gushes out of risers as they flood fields further down the terrace.

Every autumn I usually pluck a plant from a field for her to bring to school. I don't know how much her classmates learn from having such things in their classrooms, but I know she takes pride in the plants, and, of course, in being the Promethean bringer of such things. One year, vindication for my pulling of plants came when, as my daughter clutched a soybean plant with its roots wrapped in a cone of newspaper, a fellow parent turned to us and cried out, "Oh, look, edamame!" Edamame is, of course, the Japanese name for green, immature soybeans still in their pods—the pods turn yellow and then tan as the beans within ripen—as well as the dish of steamed or boiled beans served in their pods, which has become popular in the United States. That we could be surrounded by thousands upon thousands of acres of soybeans but only know them through encounters in Japanese restaurants or from glimpsing frozen bags stacked in glass-doored cases in health-food stores saddened me for a moment, but it also made me realize how important it was to get back in the field, to bridge the gap between my fellow urbanites and the farmers upon whom we depend.

Rice and soybeans are often grown in rotation, which produces a couple of advantages. First, soybeans help to return nitrogen to the ground, which rice, like all grasses, consumes freely. Second, as a dry-field crop, soybeans help break up possible outbreaks of pests, disease, or weeds that would, if given the opportunity, readily build up in a one-crop system. The other popular rotation is crawfish, which allows farmers to make good use of the rice stubble as a harborage for the crawfish and to avoid having to pull down levees. (Once upon a time, some farmers would simply leave fields fallow in nonproduction years, but that is a difficult option for farmers who are often under pressure to produce returns not only for themselves but also for the landowners with whom they share any profit.) Soybeans and rice share a considerable history. Both were first cultivated in Asia and then enjoyed a dispersal. Both in antiquity and its introduction to the Americas, soybean was the latecomer. Rice always seems to come first, and, given its prominence as the foundation of the diet of more than half the world's citizens, that is probably as it should be.

Rice first came to the Americas in 1685, when a severe Atlantic storm—possibly a hurricane—drove a brigantine bound from Madagascar to Europe into Charleston's harbor. While the ship underwent repair, its captain passed along some seed rice to a local doctor, Henry Woodward, who in turn passed the seed along to some of his friends, who were able to grow the Madagascar rice, adding another crop to American agriculture.[24] When plantation owners found that some of their slaves knew how to grow rice, it wasn't long before demand for such slaves reached Africa. It is quite possible that some of the dispossessed Acadians experienced these Carolinian plantations, with their complex networks of levees that took advantage of the ability of ocean tides, by raising the water level, to push freshwater into fields where it could be held when the tide waned. Such a sight would, as the historian Carl Brasseaux has pointed out, certainly have reminded the Acadians of the similar structures they had built to reclaim land in the farms they had been forced to leave behind in what would become Nova Scotia.

Meanwhile, in Louisiana, the city of New Orleans was founded in 1718 and the first slave ship arrived only one year later. A few records point to slaves with knowledge of rice being sought out, and the general sense is that rice was seen as a crop that could be grown where nothing else would. Its principal market was, so far as I can tell, as fodder for slaves. The rest of the state's residents were much more focused on trying to grow wheat for bread or learning to master the many things one can make with the continent's own grain, corn—indeed, our first accounts of gumbo are of the dish being eaten not over rice but over cornmeal mush, or *coush coush,* as it came to be known. (The grain is American; its preparation as a mush is European; and its name is African, taken, scholars believe, from its resemblance to couscous.)

Rice remained an opportunistic crop for much of Louisiana's early history. Several observers noted its presence among Acadian farmers, especially among those who had moved away from the Mississippi River and its immediate tributaries to begin to populate the prairies west of the great Atchafalaya Swamp. Among the prairies and bayous, small farmers planted patches of "providence" rice, broadcasting seed in low-lying spots—in coulees and ponds, according to Lauren Post—where the grain might take advantage of its natural tolerance for standing water. Although the supply was never great, it did cultivate, as it were, a taste for the grain that may have been as important as anything else, keeping rice growing an active part of the region and thus an active part of the landscape.

Except for a brief moment immediately before and after the Civil War, rice was never subject to large-scale agricultural efforts. For about twenty years, between the 1850s and the 1870s, the availability of steam power to pump water into fields and a depressed market for sugar made rice agriculture an interesting proposition for river plantation owners. But, like the Carolinian planters, they suffered from the loss of enslaved labor after the Civil War, and, given the higher costs of planting, when people actually had to be paid to work, it only took a few years of returned profitability in the sugar market for planters to return to raising cane.

Without a doubt, what happened next is astonishing not only for how it happened but also how quickly it happened, with people and technology converging, in a wholly intentional way as it turns out, to transform what was a pastoral landscape into perhaps the most modern landscape Louisiana had ever known.

What opened the door, quite literally, to the transformation of the Louisiana prairies into an agricultural landscape was the completion of the Louisiana Western Railroad in 1881, which bridged the gap that had existed since before the Civil War, when one line had reached as far west as Berwick from New Orleans and another line as far east as Orange from Houston. The land through which the railroad ran had been, from the point of view of the powers that be, sparsely populated, filled mostly with grazing cattle. Several companies sprang up to fill this newly opened, and fertile, landscape, which was advertised as "free from protracted droughts which afflict Kansas and other prairie regions."[25] They bought up large chunks of land and then sold it in smaller chunks, often in quarter sections of 160 acres, to farmers who often came down in tours, sponsored by the companies or by the railroad (now the Southern Pacific), which was anxious to have the land settled and productive.

One company, the North American Land and Timber Company, provided free passage so that farmers could view the land for themselves as well as attend agricultural demonstrations during the tour—they also offered financing if a deal was struck.[26] North American preferred to sell only one-quarter out of a section because the new tenant would improve their chances of selling the other sections at a better price. They were not stingy in their efforts, however, contributing land for rights-of-way and digging canals with steam dredges to improve drainage. Jabez Bunting Watkins, who had once been a farm mortgage broker in Kansas but who had found his true calling in leading the efforts of North American to settle the Louisiana prairies, once proudly observed that he was responsible for the importation of more than one hundred tons of farm machinery, much of which came from faraway England. He intended to "surprise the natives. . . . We have gang plows that will turn 50 acres a day."[27]

Thanks to such efforts, the population of the Louisiana prairies nearly doubled in the last two decades of the nineteenth century, from 126,000 to 240,000. As the historian Henry Dethloff noted, "Most of the new settlers were farmers from the North and Midwest lured by cheap land and driven by droughts and blizzards that beset the midwestern prairies in the 1880s and climaxed with the terrible winter of 1886–87." These farmers were used to growing crops such as wheat, which had never done well in the long, hot, and humid Louisiana summers, but it did not take them long to turn all the equipment they had brought with them—steam tractors, harrows, plows, and threshers—to the task of growing rice.[28]

What they discovered was a landscape practically made not only for holding water but also for working with heavy equipment: under the Louisiana prairies lies an impervious claypan subsoil that holds water like cement and supports heavy equipment that would otherwise bog in deeper soils. Combined with a growing season that typically accommodated both a first crop and a ratoon, or second, crop and abundant rainfall, what had

RICE PLANT

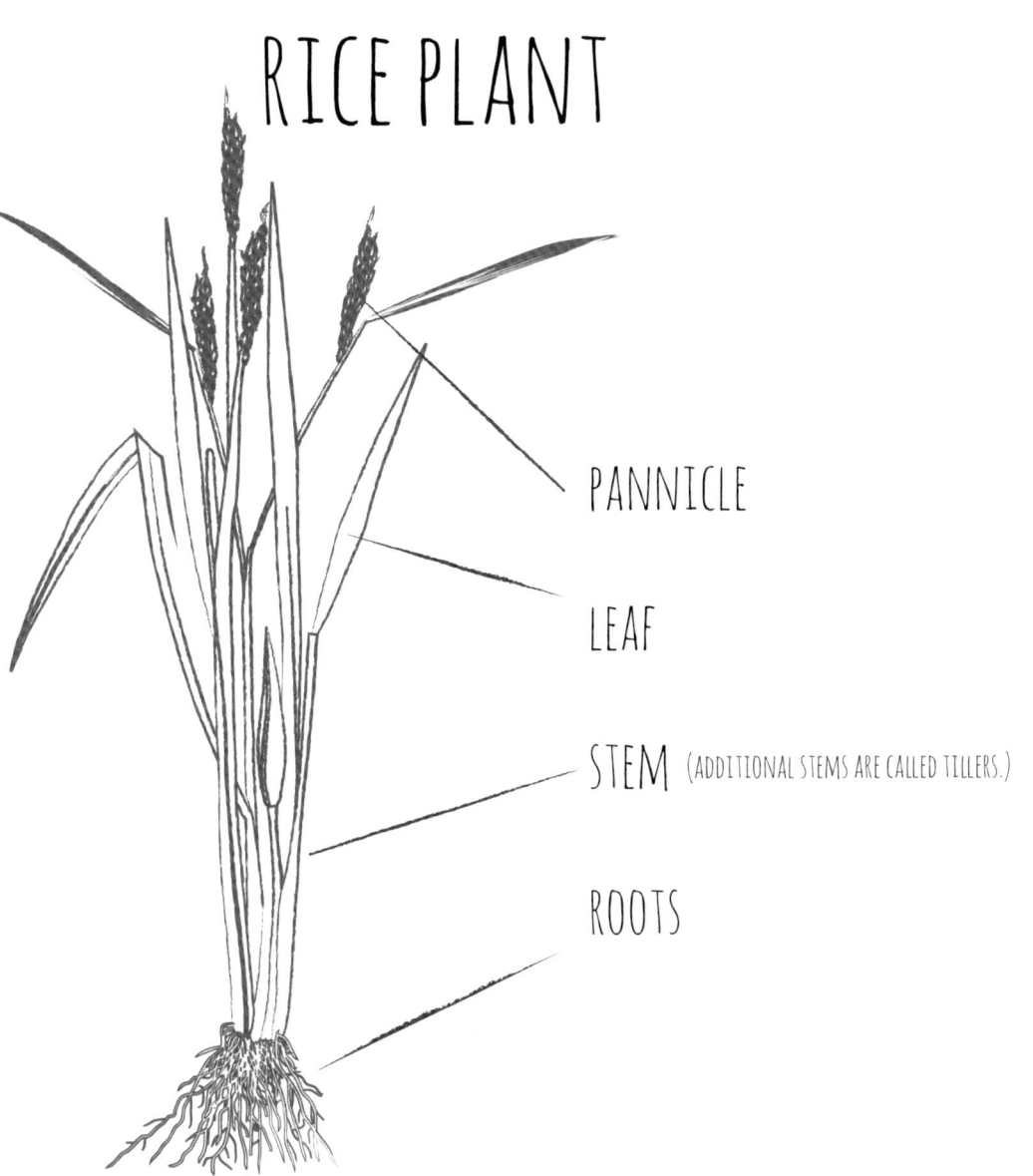

PANNICLE

LEAF

STEM (ADDITIONAL STEMS ARE CALLED TILLERS.)

ROOTS

The parts of a rice plant.

once been small-scale agriculture intended for local consumption became an agricultural powerhouse whose output more or less doubled each decade during its first forty years: 834,111 bushels in 1879 became 2,721,059 in 1889, then 6,213,397 in 1899, followed by 10,839,973 in 1909, and 16,011,607 in 1919. The scope of the mechanization can be gleaned from equipment shipments during these years: the Southern Pacifica Railroad shipped one twine binder to southwest Louisiana in 1884, two hundred in 1887, and one thousand in 1890.[29]

Things like twine binders were not passively consumed; as a number of historians have pointed out, it was Maurice Brien, himself a midwestern transplant, who figured out how to use a wheat-harvesting twine binder for rice harvesting. It took Brien a few years, but by 1886 he had a working solution, which contributed to rice becoming a commercial crop for southwest Louisiana. As Lawson Babineaux noted, "This modernization of rice production led to a revolution in agriculture on the prairie which in turn was responsible for the establishment of its rice industry. In place of the hoe or walking plow, the farmers from the North used their gang plows. They used seeders and disc harrows instead of sowing by hand."[30] Within a mere decade, Louisiana went from third place in rice production to the leading producer in the United States. Buoyed by a steady stream of immigrants from the north and new European immigrants, particularly the Germans of the Rhine Palatinate, as well as a steady stream of often locally developed technological improvements, rice became the only Louisiana crop to see growth in production in the decades that followed the Civil War and marched toward the twentieth century.

It is worth noting, I think, that not only is the swiftness of the transformation remarkable even today, but contemporary observers found it equally extraordinary. Speaking in 1889, Louisiana's secretary of agriculture James Wilson waxed rather rhetorically about the town of Jennings: "This is not a typical Southern town I know. If you were to drop a Northern man down here in the night, when daylight came, he would say, 'Well, I happened to fall upon a Northern village. Everything looks so much like that what I have seen in the North. Those homes look like Northern homes. Those people look like Northern people.'"[31] A few years later, writing in as dull a venue as the *Biennial Report of the Louisiana Commissioner of Agriculture*, T. S. Adams observed, "Improved rice seeders, self-binding reapers and steam threshers are frequent silhouettes upon the sky-skirting prairies, declarations of Yankee invasion of the once peaceful abode of the Attakapas."[32]

~~~~~~~~

Reading histories of early efforts by the railroads and other private investment companies to settle the Louisiana prairies, you realize that the quarter sections of land that were being marketed to prospective buyers were being sold on the very basis of their mechanizability. It's not clear if the northern and European farmers necessarily expected such machines to be a part of the landscape or if it was part of the overall vision of progress being advertised, but Watkins, working for the North American Land and Timber Company, and others regularly featured steam tractors plowing and cultivating as part of their tours.

Contemporary memories of twentieth-century developments suggest that mechanization, or at least a desire to mechanize, was a significant part of farming practice.[33] Although not that many farmers began with steam tractors, it remained a desire to have one. Most of the early agricultural machines—whether gang plows or threshers or bailers—were pulled by horses. Self-binding reapers were drawn by rather large mule teams, with the machinery driven by a bull wheel that created the mechanical energy required to first cut the rice stalks, then push them onto a conveyor of some kind, and finally bind a group of stalks into a shock, which was dropped to one side as part of the conveyance movement.[34] The shock, of course, dates to antiquity and remained a part of agricultural practice until harvesting and threshing were combined into one machine, aptly named the *combine,* perhaps one of the greatest labor-saving devices ever invented.

Both the tractor and the combine did not really come into their own in Louisiana until the 1950s, when small gasoline-powered tractors became affordable and combines became self-propelled. The oldest farmers at work today have memories from that time, and many of them recall fondly their family's first tractor or combine. Until then, horses remained a significant part of farm life in south Louisiana, except on the largest or most profitable farms. The rise in productivity brought about by the introduction of these machines, as well as improved varieties (known as *cultivars*) and synthetic herbicides, can be glimpsed in the history of rice yields: after the precipitous rise in the last two decades of the nineteenth century, yields hovered around 1,800 to 2,000 pounds per acre throughout the first half of the twentieth century. Only after the Second World War does rice production rise significantly, and then it does so with pent-up ferocity, doubling to more than four thousand pounds per acre by the late sixties. Further rises occurred in the eighties and again in the late nineties thanks to the development of cultivars, many of which were the product of the Louisiana State University (LSU) AgCenter's Rice Research Station located in rice country itself, between the towns of Crowley and Rayne. (Such is their faith in technology, Louisiana rice farmers tax themselves in order to fund the research, and the relationship between the station faculty and staff and area farmers is quite open and warm.)

One should not imagine that such modern machines simply appear magically on the landscape and everything changes. Rather, they first appear in the fields of more prosperous, or more adventurous, farmers. Others stop and stand at the edge of the fields, observe, and wonder if it's worth the investment. There is a great deal of conversation in feed and seed stores, in hardware stores, at the mills, and at church. How is it? What's it like to work? Does it increase yields? Does it damage the land any? The questions are direct. These are men and women who have, thanks to mechanization, always worked alone or in small groups of two or three. As the years pressed on, they worked more and more acres, keeping the same amount of people per farm, resulting in a decrease of the ratio of people to acres year by year. (Even the need for day laborers, once a space and time within which blacks and whites worked together, would slowly wear away, and only while in town would white farmers encounter black city dwellers, and they would have little in common and little to discuss.)

The first tractors were small and quite simple, barely more than a motor harnessed to a pair of large wheels, but they could, unlike horses, work hours on end and, perhaps as important, they were more than simply machines for pulling plows and other devices through a field: they were also portable power plants. The first tractors came with flywheels on one side of the engine around which an operator could wrap a belt to drive another machine. One of the primary uses for such a power hookup was to drive the early threshing machines, which were pulled to a spot in a field and the shocked rice brought to them. Later, the flywheel became the power take-off, or PTO, spindle. The PTO remains a part of most modern tractors, but its importance is being increasingly supplanted by hydraulic hookups that allow a machine to be powered via the tractor's own system. Remote hydraulic power enjoys a number of advantages: it is closed to human encounter save at the point where the power is delivered and it does not require a complex system of joints and axles, which are more prone to failure and snaring a wayward hand or foot.

As the machines they used became more capable of driving other machines, farmers began to seek out devices that solved particular problems. Sometimes they worked on their own or sometimes they sought out the help of a shop. No matter the origin, if something worked it would soon find its way onto other farms, with resultant variations as by-products of particular needs or ideas or experiences. Out of such moments were born PTO pumps, for moving water out of canals and ditches, and PTO ditchers, for creating the shallow, wall-less traces that allow water to drain out of near-level rice fields, as well as side plows and water levelers and myriad other devices and machines.

All of these things, some freshly painted and some careworn, allow farmers to transform the gently sloping landscape of south Louisiana into a series of carefully calculated, as-close-to-level-as-you-can-get rice fields that they maintain all year round by pulling levees up and then pulling them down, by mucking about inside flooded fields with giant tractors in order to pull down hills and pull up holes to make as-close-to-level-as-you-can-get cuts. There is never anything perfect. There is never anything finished, especially when confronted with both the uncertainty of cultivating living things while at the mercy of the randomness of the weather, a randomness redoubled by the greed of the commodity markets, which are more than happy to punish mistakes and successes alike.

In the face of such forces, the ability to get a side plow made just for your tractor or a boat that works best in the kinds of fields you have seems a small affordance of security, a slight hedge against the many risks. It is no wonder then that the ties between farmers and fabricators are so strong. The men in the shops have, working as they do across farms, the knowledge of what can, or will, go wrong, and can right that wrong the next time they make it. They are not invested in mass production, and they do not have to wait to retool an entire manufacturing line. Whereas robots and jigs may require weeks, or in some cases months, to be changed, men working with hand tools can change immediately. The next weld will be twice as long, and the next weld is not next week, or even tomorrow, but right now.

The same immediacy, and the same knowledge, applies to repairs. A farmer like Jerry Leonards comes into a shop in need of a repair. For him, it may be the first time this piece of gear has broken, but in most instances it is not the first time the fabricator has seen that piece of gear broken. A great deal of the gear that farmers use is built by large national or international companies that are, in effect, making general-purpose machines and equipment. They cannot know the particular uses and abuses that their products will encounter in the fields. Farmers in Louisiana in particular feel that that is the case for them: a lot of the gear they use is made with the corn and wheat fields of the Midwest and plains in mind. Rice is rough, they say. Its hull can wear almost anything down more quickly than any manufacturer has ever anticipated. It is up to the men in the nearby shops to know how to put things right, often improving on things as they go.

This sense of collaboration between farmers and fabricators within the sphere of rice production is reinforced to some degree by the bounded nature of rice agriculture in Louisiana, which is concentrated in a handful of parishes among the prairies—there is another rice-growing district in the state, in the northeast corner, but it is really a part of the Arkansas–Mississippi Delta region.[35] There is, then, somewhat of a sense among rice farmers of being an island of rice production, a feature they argue, especially when they are feeling grumpy, of the "land not being good for much else." Fortunately, they note, the feature of the land that makes it so poor for other crops—the relative shallowness of the topsoil and the ready access to the claypan subsoil—also allows them to operate machinery where it would not, in deeper soils, be possible. For every curse, from a farmer's perspective, there is a somewhat smaller blessing.

~~~~~~~

Rice planting begins a year and a half ahead of any seed being dropped onto the ground. It begins in the previous harvest, as farmers make mental notes about how much a field produced, what variety they planted, and how much fertilizer and/or herbicide and/or pesticide they had to use; each of these adds costs to a field that must be subtracted from its profitability. These notes wrap up a season's worth of observations about how well a field flooded and held water, how even the growth was, and whether there were uneven patches and whether those patches resolved themselves or could be heard, or felt, in the cab of the combine during harvest. All of these details are appended to information the farmer already knows about the field, often from his own previous experience but sometimes from other farmers or from landowners who know a field's history.

What emerges from this compilation is an action plan: things that must be done in the next eighteen months. Such a plan must take into account not only this time lag but also what will happen in those intervening months, for the field will either be rotated into soybeans or crawfish or it will be allowed to lie fallow. If soybeans, then there are levees to be pulled down. If crawfish, then there are drains to be checked and ramps to be created so that crawfish boats can slide into and out of fields without damaging levees. If fallow, there is always plowing to be done, turning weeds into fertilizer and smoothing the ground.

Sixteen months after harvest, the work begins anew with the leveling of cuts. In some cases, this is merely maintenance, and a farmer will only spend an hour or so in each cut, adjusting for shifts in the soil or nuances of the land's topography that he slowly learns over his, and its, lifetime. In other cases, there may be a significant reworking of a field planned, usually resulting in either fewer cuts or in straighter lines, if possible. Fewer cuts increase the amount of arable land, reduce the number of levees—each of which must be maintained—and accommodate the increasing size of equipment. Combines in particular have grown steadily in size, and their bigger headers make working smaller cuts increasingly difficult.

When all the intervening steps have been taken, it is time to plant the rice.[36] Perhaps what farmers worry about most is the temperature. After all, they have carefully groomed the landscape, controlling, as best they can, one dimension of nature, and they are prepared to spend money on herbicides and pesticides to control another dimension of nature, and they already plan to flood the fields, controlling yet another dimension. But they cannot control the temperature; in fact, if it is below fifty degrees Fahrenheit (ten degrees Celsius), rice shows little interest in germinating. Most farmers choose to wait until the average temperature is well into the mid-sixties or more. Given that early planting also exposes seeds not only to the cold but also to migratory birds happy to feast on an easy meal, one might imagine that farmers would rather plant as late as possible, but doing so risks not only lower yields but also increases the plant's exposure to bugs and disease. Like everything else in farming, there is no clear-cut answer, and each farmer answers the question based on his own predilections and experiences and exposes himself to the scrutiny of his neighbors and colleagues. Guessing rightly, which is sometimes the same as getting lucky, nets you praise, and perhaps some envy. Being unlucky more often than not gets you a sympathetic head shake: almost everyone recognizes that "there but for luck go I."

How one seeds a field varies, and it comes in two main forms, wet or dry. Wet seeding has been the preferred method on the prairies for some time now, and it involves flooding a field for a few days before having the seed dropped from a crop duster. Wet seeding is one way to control red rice, a weed that is the bane of a rice farmer's existence—because it is so closely related to rice, it makes it difficult to treat with herbicides—and it is also convenient for farmers who rotate rice and crawfish. When wet seeding, most farmers give the seeds time to "peg" (send a small root down to anchor the seed) and then drain the field for a short time.

In the case of dry seeding, the seed can be broadcast from the air, as it is when wet seeding, or it can be dropped onto the ground using a seed drill, a process known simply as "drilling." How long the seed is allowed to develop in a field that is flooded depends a lot on the year's rain. Interestingly, not every farmer wants rain during this time. Some, in fact, will even drain a field flooded by rain and pay the fuel bill to flood it with well water, believing that the mineral-rich groundwater makes for a better crop.

No matter whether a field was wet or dry seeded, or flushed by irrigation or by rainfall, at some point it is flooded for the rest of the growing season, sometimes called the "permanent flood." During the months of May, June, and much of July, farmers anxiously keep tabs on their fields by driving along highways and field roads, surveying for uneven plant development or any signs of disease or pests. As the crop gets thicker and taller, the levees disappear, and will only really be seen again when the rice ripens to a golden-tan color and the levees remain green with weeds.

If a farmer has been lucky and everything has gone according to plan, then he can begin to get combines and grain carts and grain trucks ready to harvest. Grain bins are inspected for any equipment in need of repair or replacement, and market prices are checked with greater frequency. Mills are called, and some rice gets "booked" (sold in anticipation of harvest). A happy resolution to months of work, and hope, seems at hand, one that cannot be disturbed even by the traces of wild pigs rolling and rooting in the fields that line bayous and coulees, their preferred travel network. But July and August are prime months for hurricanes and tropical storms, which can lay standing rice down with their strong winds and torrential downpours. I was in the field with one farmer outside Lake Arthur as Hurricane Katrina approached the Louisiana coastline, and even as it became apparent that the storm was going to strike further east, he could ill afford to take the chance of a sudden turn laying low, quite literally, a year and a half's work. He climbed into his combine and, with his brother and father, began to harvest as if there were no tomorrow, because there was a chance there would not be one for the rice.

If no storm approaches and the crop can be gathered, it is generally a pretty spirited time. There is perhaps no happier being than a farmer riding in a combine, thrilling to the sound of rice as it tumbles through the machine's grain paths and fills the bin behind the cab. There is immense satisfaction in sitting in the cab, riding just behind and above the header, and watching the reel push the rice stalks, their heads drooping with rice, into the cutter bar. Kernels, stalks, and leaves fall into an augur or onto a moving conveyor belt called a "draper" that pushes everything toward the combine's throat, where it disappears into the belly of the machine. Once inside, stones and other heavy debris are diverted away from the threshing drum, where the rice kernels drop down through a series of sieves before being lifted up into the grain bin. The straw left behind by the threshing process is walked carefully back out of the machine, being shaken as it goes to glean every possible bit of grain. Eventually it is cast out the back of the machine, where in most instances it is chopped and strewn as organic material being returned to the earth.

If the weather has been dry, then the combines ride pretty easily on top of the ground. If the weather has been wet, then they tend to slop through the fields, making a mess but not making things too difficult. There is an in-between state, where the fields are wet enough that the combine's tires slip through the topsoil but dry enough that the mud sticks to the tires, a situation that most farmers dread: fuel costs are greater, as is the wear on men and machines—working a combine means constantly turning inside a field, a task made difficult when the field itself wants to stick to anything

A rice field levee cut open and ready to have a flexible pipe drain installed.

with which it comes into contact. This state of a field also tends to dry that way, leaving ruts that will be hard on both human bodies and metal frames when it comes time either to cultivate for soybeans or, if things go well, harvest a second, ratoon, crop.

First crops are usually harvested during late July and late August, but I have seen combines in the field as early as mid-July and as late as mid-September. Like everything else in farming, it all depends. And for farmers, such dependencies don't end in harvest, with delivery to mills requiring them to negotiate not only the price of the grain, but also its quality: any price you see for a commodity is only the starting point for a negotiation, always downward, for a final price. Mills judge rice based on an optimum water content, how fragile the kernels are, and how much weed seed and other undesirable materials are mixed in with the rice. Mills and brokers are in turn looking to sell the rice to food distributors or packaging plants, and, in a global marketplace, the price of rice rises and falls on the overall supply and demand, which is as much about politics and perception as it is about economics and ecology.

ANTECEDENTS AND PARALLELS

In 1878, on the eve of the transformation of the Louisiana prairies into rice fields, the weekly French newspaper in Saint John the Baptist Parish, *Le Meschacébé* (The Mississippi), printed the last of a series of folktales attributed to a Pierrot, who readers then and now have to assume was African American, both because the title of the series was "Contes Nègres" and because, unlike the rest of the paper, the story appeared to have been told in French Creole.[37] The story tells of a young carpenter who, having fallen in love with a princess, rises to the challenge her father, the king, has set: to build *ein bato ki té navigué on la terre com on la mer* (a boat that can go on both land and sea).

> Once there was a king; he and his queen had just one girl, who was pretty—but pretty! Prettiness could go no further.
>
> All the sons of the other kings in the neighborhood, as well as the generals, colonels and corporals all wanted to marry her. But this king didn't want that, no indeed! He didn't want to hear any arguments that would separate him from his daughter. But I'll tell you-all, I believe the daughter herself did want to leave home. There was a young carpenter who worked on the king's house, who was a much finer fellow than all those princes. He had already admired the princess, and she had looked him over too. Then they went to making sweet-eyes—no joking about it and then, you hear me, they got secretly engaged.
>
> People were after the king and queen so much that he put a notice in the paper to say that he would give his daughter to the one that could make him a gift of a boat that would go on both land and sea. All the princes and their young companions got tool boxes, ran into the woods, and started chopping trees to build some boats.
>
> When they were all working harder than horses, an old one-eyed black man came into the woods, asking for charity. He had nothing on but old

rags, which stank worse than an alligator. Those princes held their noses and said, "Pooh! On your way, you old crocodile!"

The young carpenter had stopped work and was about to eat his dinner when the old man came up to him: "Hey, old grandpaw, you got here just in time. Sit down on my logs and we'll eat together." When they had finished, the old man took out a clay pipe from his pocket and started to smoke. He blew out a big cloud of smoke and it smelled of roses, jasmine and violets. A wind then started to blow and—Vroup! it drove away all the smoke. Then what do you think the carpenter saw? Instead of the old man, there was a pretty fairy seated in a gilt buggy drawn by pigeons. She had a diamond wand in her hand and she said to the young carpenter, "My son, you have a good heart, better than any young man in this district, so you will get what you deserve."

She tapped the wood with her wand and it turned into a boat, a finer one you-all never saw, with a crew and all you need to sail off. Then she introduced the ship's officers to the young carpenter. "This one is a good marksman. He can fire anywhere and hit anything. This other one is 'Good Diver.' He can dive into the sea and bring up anything you want. The next is 'Good Runner,' who goes like the wind and can circle the world like a telegraph. This other one is 'Good Listener,' who puts his ear to the ground and hears all that's going on everywhere. The other members of the crew don't have to receive orders, they can guess whatever you want done and wherever you want to go. With these you will win the king's daughter."

The fairy went off into the air and the young carpenter got on board and blew the whistle the fairy had given him. They immediately set off to meet the king. Once arrived, everyone came on board to try out the ship. But the king was sad at the thought of giving up his daughter, and when they reached the sea he told his wife to throw her diamond ring into the water. Having thrown it, she began to cry and say she would never give up her daughter unless someone returned her ring. This made the young carpenter feel awful, but Good Diver came and told him he was able to recover the ring immediately. He jumped in, boudjoum! like a big bullfrog, and in five minutes returned the ring to the queen. Then they turned the ship about and the carpenter went to talk to the king. The king declared, "I have another thing to request before I give you my daughter. The queen wants the feathers of a bird of paradise, so go hunting and kill the bird, then bring it to me, and then you will have my daughter."

This saddened the young carpenter, but before he had a moment to think about it Good Marksman said, "That's just my kind of business." He called Good Runner and went outside. He aimed carefully toward the land of Paradise and fired. Boom! Then Good Runner left at a run to fetch the bird, but Paradise was so far off that Good Runner grew tired and lay down to sleep. Everybody waited, but Good Runner did not come. Good Listener put his ear to the ground and he heard Good Runner snoring a long way off.

He called Good Marksman, who put a little bomb into his gun and aimed it near Good Runner's ear. The bomb went, "Pralapaow!" and woke Good Runner and he galloped back with the bird of paradise.

The king had nothing left to say, so the young carpenter married that lovely girl, and they had a great wedding, with everything handsome. They would caress each other; they billed and cooed like pigeons going, "Roucoutoucou." Then they had many children, and lived as happy as fishes in the water. My friends, charity always pays, let me tell you! (Reinecke 1994, 22–25)

Many readers will recognize the tale's Europeans roots, which are best seen in the magical helpers. The helpers, in fact, are such favorites that in another version of the story in Louisiana, they are the central focus and the boat is a bit of a side note. Across Europe, many versions of the tale feature only the helpers: some readers may recall the tall tales often associated with Baron von Münchhausen, and, indeed, it is not uncommon for various of the helpers to turn up in other tall tales.[38] How it got to Louisiana is not clear, but it is very close in form to both the German and French versions of the tale, both of whom were present along this part of the Mississippi. Other tales recalled by Pierrot, and printed in the paper as "Contes Nègres," were, according to folklorist George Reinecke (1994), more clearly African in origin.

In a number of the most renowned versions of this tale, the protagonist is the youngest, and/or dumbest, of three sons.[39] The two older brothers miss their opportunity to have the ship by mistreating the magical helper, almost always an old man or old woman who appears as a beggar or a vagrant in ragged clothes. In this version of the story, there are no brothers, only princes, with no relation to our protagonist, who is simply a carpenter. Reinecke speculates that some of the allure of the tale, for its inferred teller and audiences, was that a carpenter would have been part of most plantation estates and would have likely been either an enslaved African or a person of color with a potentially complex status in the antebellum period. At any rate, in the decades following the Civil War in which this tale occurs, we have at the very least the crossing of class boundaries in having our princess already in love with a craftsman.

Our craftsman has, in comparison to the princes in the tale, at least some familiarity with what it takes to build a boat, although it is not his skill that produces the boat, sadly for our purposes here. Instead, it is his willingness not to reprove others and to share what he has, which is similar to most versions of the story. Generosity extended, the old one-eyed black man who "stank worse than an alligator" transforms into a pretty fairy, in her own means of conveyance, "a gilt buggy drawn by pigeons." It is the fairy who turns the wood into a boat, and who then staffs it with helpers: Good Diver, Good Runner, Good Listener, and Good Marksman— the last of which is not introduced at first but turns out to have been in the boat as well when the time comes. What follows is not an elaborate series of scenes, each featuring a particular helper, but rather only two scenes, one that features Good Diver and one that features the other three. (The combination of Good Runner, Good Listener, and Good Marksman is quite common.)

The first scene, however, features a character not commonly glimpsed in other versions of the tale, the Good Diver. Having been given the ship and his companions by the fairy, who leaves him with the words "with these you will win the king's daughter," the carpenter sets off to find the king and show him the boat. Everyone piles on board to try out the boat, and the king, saddened at the thought of giving up his daughter—a lovely human touch in the middle of the tale—tells his wife to throw her diamond ring into the water. Good Diver is there, of course, to recover the ring and to keep the story on track, and what he gives us is a doubling of the amphibious in the story, something made quite clear when the narrator tells us "he jumped in, boudjoum! like a big bullfrog." The emphasis on the amphibian appears again when, at the end of the tale, the carpenter and princess were happily married and "had many children, and lived as happy as fishes in the water."

The mixed nature of the Louisiana landscape is explored elsewhere in Louisiana folklore, though interestingly, it is mostly featured in tall tales and legends, forms in which fact and fiction are themselves confused, pressing listeners to consider what could be real and what must be made up as a tale unfolds. The folklorist C. Renée Harvison collected a compelling tall tale from Pierre Daigle of Church Point that later appeared in *Swapping Stories*, the volume she edited with Carl Lindahl and Maida Owens:

My grandfather told a lot of tall tales, and we had an uncle who would come. He was a bum. He traveled all over. He lived here a while until they kicked him out and then he'd go somewhere else. And he'd spend several nights at the house during the wintertime. I was very happy to see him because he was my television set long before television was invented. We'd stay up all night and tell stories about buried treasure and that kind of thing. Those were not folktales. They were just lies he made up.

Tall tales—we had several. My grandfather used to tell one to all the grandkids, and that one stuck to my mind. He said—this was all said in French, but I'll tell it to you in English. "One morning, I'd gone hunting." Of course he had an old musket, muzzle-loading musket.

And he said, "I had my horn, my powder horn hanging from my belt, and my lead ball pouch hanging on the other side." And he said, "I was way out there in the forest, and I shot something." And he said, "When I came to reload, I noticed the pouch had broken open and all the lead balls had fallen out. There I was way in the forest, plenty of powder and caps to load the gun, but no lead balls."

So he said, "I started back home. I came to this small lake in the forest. On the bank of that lake was the biggest alligator"—which makes me think this one originated here because alligators were not called alligators, they called them *cocodrie*. They thought they were crocodiles. Anyway, he said, "The biggest alligator was sleeping on the banks. Boy, I sure would like to kill that alligator for the skin! I could skin it and sell the skin." He said, "I started thinking what I could use for musket balls. That morning, before going hunting, I passed by a peach tree, and I'd plucked some peaches and I'd eaten them, but I'd put the seeds in my pocket when I came back." He said, "I decided, maybe a peach seed. I packed that powder in, put a peach stone, I packed it in, went right up to that sleeping alligator and shot it right between the eyes. That thing pulled up, rolled over, crawled back in the lake and disappeared. Well, I didn't get him."

He said, "Five years later, I went hunting in that same place. Right on the banks of this lake there was a beautiful peach tree growing. It was loaded with peaches. When I ran out to grab a peach, it swam out in the middle." He said that peach [stone] had sprouted in that alligator's head and made a beautiful peach tree! (218–19)[40]

Curiously, the tree at the edge of a lake or pond also features in a number of treasure tales from the area. One legend told by Samuel Gautreaux, and recorded by Barry Jean Ancelet, begins with just such a figure: "In Charenton, north of Lake Charenton, there was an old Indian named Jim. And they asked old Jim where might be a certain marked oak tree north of the lake" (1994, 147). In another story, told by Stanislaus Faul of Cankton, the water becomes a magic well with buckets that rise and fall unattended and is marked by an oak tree that has been moved to mark the place (1994, 150–51). Another tale, collected by Lindahl and Owens, but not published in *Swapping Stories*, also imagines that the land and water hybrid is central to treasure:

Her . . . father one time, when he was a boy, there used to be a store just west of here, Field's. It's still there. He was playing there on the porch, and he had a nugget. One of what they call a drummer, a salesman, come by and said, "Let me see that, son." He looked at it and it was supposedly silver. He gave him fifty cents. Told him, "I'll pay you more if you take me"—now, there's a little creek out there, Windham Creek—"If you take me back and show me where you found this." He said he was just about eight or nine years old, playing along this creek. He was sure he could find it, so he took this fellow up there to show him where it was. He never could find the place where he found it. One time out east of town here, a guy found a stump full of money. A hollow tree. Come to find out, it was slot machines stolen from the old Pines nightclub. All these nickels and stuff was hidden in this stump, in this hollow tree.

Whether it is a hollow tree along a creek that contains money from stolen slot machines or a pirate in a tree demanding a drink, the trope is well established in the folk culture of the area, but, as I noted previously, it is mostly relegated to the realm of tall tales and legends that ask us either across an entire small text or at individual turns within a larger text to consider what is the nature of reality.[41]

Folklorists, among others, have long considered the relationships between the geography of a region and the lore. Sometimes the relationship is an obvious one: the landscape has this feature, and so it will figure in the lore, or the residents of the area perform this action, made possible by this landscape, and so this kind of knowledge, or stories about these kinds of actions, are part of their traditional repertoire. Mary Hufford's analysis of fox hunters in New Jersey (1992) is perhaps the most striking in this regard, as well as Erika Brady's examination of trappers in Missouri (1990).

As these studies suggest, the relationship between a region's natural landscape and the human cultural network of ideas can be tightly integrated, but if my investigation into the relationship between actual landscapes of land and water and imagined landscapes are any indication, then we should be careful about assuming that the landscape of a region permeates its folk culture through and through. It does not. In particular, it does not spread evenly across the repertoire of traditions available to practitioners—just as not all traditional practices are available to all individuals within the group that hosts those practices. As glimpsed above, considerations of the intersection of land and water are largely restricted, within the realm of oral tradition, to legends and tall tales. That is not to say that neither land nor water appear elsewhere or that they do not appear in conjunction, but the intersection, or interaction, of the two does not play, typically, a significant role within the larger thrust of the story, the way it does in, say, the legend of treasure hidden in a hollow tree standing by, or in, a body of water.

I once spent several hours with Barry Jean Ancelet as he went through what is perhaps the most extensive collections of Louisiana folklore to date—the index for which he keeps in his head. In the end, the list of texts of possible interest was rather small. Such an outcome should not be taken as an indifference to the landscape, but rather, as Gay Gomez suggests in *A Wetland Biography* (1998), that we need to widen the archaeological record; that is, people do not necessarily and always think in words. Sometimes they think in actions, and sometimes they think in things. This might mean a variety of things, but in the case of Louisiana, it means that agricultural impulses to flood up prairies, plant them with rice, and then hold the water in order to grow crawfish must be taken as important, and as traditional, an act of mentation as the telling of a story.

~~~~~~

If we turn to artifacts, then it is clear that the residents of south Louisiana have long made, or at least thought about, boats that could handle the mixed nature of the landscape on which they found themselves, and, perhaps, there is no more famous of such an artifact than the pirogue (usually pronounced either *pee-rogue,* or, with the hard *g* dropped, *pee-row*). With its origin in the Americas lost in prehistory, the pirogue is, quite literally, the first boat of Louisiana, and long popular for its ability, as anthropologist Ray Brassieur noted, "to glide on the dew."[42] Such an image, of a boat gliding across a field still, and only, graced by the morning dew, is remarkable. There are other such images to be found in Louisiana tradition, and we take up one in a moment, but the pirogue figures large in the Louisiana imaginary, especially in the heritage industries, sometimes eclipsing the discourse to such a degree that our fascination with it obscures more than it enlightens.

Most histories of folk boats in Louisiana begin with the pirogue. Both William Knipmeyer, in his original history "Folk Boats of Eastern French Louisiana" (1976), and Malcolm Comeaux, in his update of Knipmeyer (1985), begin with the original version of the pirogue, the dugout.[43] Indeed, the dugout log is not unique to America nor the New World. It seems to be, to use an old-fashioned phrasing, a universal form. That is, it is found almost everywhere humans have settled along a shore where there were also trees of sufficient size to be hollowed out. Typically, the hollowing is achieved through a two-step process that begins with removing large chunks of the log's interior through careful burning and then refining the shape and thickness of the resulting hull using some kind of scraping tool. In societies that have passed through their Iron Age, there can be a variety of tools made for just such a purpose. In societies that have not, there is typically a collection of sharp rocks or rock-faced tools to achieve that end.

Prior to the arrival of Europeans, the dugout, as Comeaux noted, was "the only true boat used by Indians in Louisiana" (1985, 164). One outcome of the pirogue's construction with the tools they had in hand, mostly fire and simple stone, was that they had incredibly thick hulls, which made them very heavy and, it seems, difficult to handle. Another was that the boats made by Native Americans had blunt ends, making them appear more like the boats common on the Louisiana waterways today.[44]

In south Louisiana, the dugout canoe was adopted by European settlers, who, at some point in time, came to refer to the craft by its Spanish name, *piragua,* which over time increasingly referred to the particular kind of flat-bottomed canoes made in the region.[45] In the latter days of the colony, in the decade leading up to Louisiana's statehood in 1812, it is not clear how often pirogues were simply referred to as *voitures* (vehicles). As one traveler, C. C. Robin, observed during his sojourn in Louisiana from 1803 to 1805: "People in this country are so accustomed to travel by water that the generic term 'voiture' (standard French for 'carriage') is always applied to a boat. If a Louisianian says to you 'I brought my voiture' [or] 'Can I give you a lift in my voiture?' he is referring to his pirogue or skiff as a Parisian using the same word would mean his coach" (Robin 1966, 101). As a coastal state with a large, extant waterway system that had the Mississippi River as a thoroughfare and the Atchafalaya Swamp as a hub, Louisiana's first navigational orientation was to water. In such a milieu, carriages become boats, and the word used for one came to refer to the latter. (Perhaps this is the first recorded blending of land crafts and watercrafts in the popular imagination.)

With the introduction of steel tools into the process, dugouts became thinner and lighter. Axes felled the cypress trees as Europeans adopted the preference of Native Americans for the lightness of the tree's wood and its natural resistance to rot. European boatbuilders brought a greater number of specialized tools to the process as well, making it possible to refine the craft's form to a greater degree: adzes scooped out the wood from the interior of the boat; augurs drilled holes so that makers could check, and recheck, the thickness of the hulls—the holes were later filled with wooden plugs hammered in tightly; and planes and other tools made it possible to slick the hull significantly.

The better construction technology also led to a French innovation: they pointed the ends of the pirogue, making it faster, easier to row, and, when used among the trees of a swamp, easier to pick one's way through a

A pirogue pulled up along the Bayou Teche.

stand of cypress trunks: a blunt bow will simply crash into a tree whereas a pointed bow will deflect to one side or the other. The pirogue was not the only craft in common usage in south Louisiana during the colonial era and in the first century of statehood. There are a myriad of other working craft, most of them made of planks and which get us closer to the foundational form of the crawfish boat.

The formative catalog of these early boats, as compiled by William Knipmeyer in the 1950s, chronicles the development of the pirogue from when it is no longer being carved out of a trunk of a tree but is instead being made by pulling planks of wood together. The turning point in history seems to be right at the onset of the twentieth century: "Most informants remember the plank pirogue as becoming important about 1910. It is not certain whether it was created at that time, or whether it was in existence previously but not yet able to compete with the dugout in popularity" (Knipmeyer 1976, 127). Unlike later histories, which suggest that the plank pirogue arises out of the ready availability of dimensional lumber being produced by the timbering of the Atchafalaya Basin, Knipmeyer seems to suggest that the plank pirogues became popular because the demand for the boats by the timber industry was so great: the boats were used to ferry out the small crews that attacked the great cypress trees of the basin with saws and axes and who then floated out with the raft of trunks.

No matter the reason for the change, the plank pirogue joined a number of other craft, also planked, that were already plying the waterways of Louisiana. One, the *esquif,* or skiff, is more like the pirogue with its curved sides reaching from pointed bow to a usually blunt stern. The others—the *chaland,* the *bateau,* and the flatboat—had blunt bows and sterns, and, in the case of the chaland, were rather a lot like many of the crawfish boats being built today.

Of all the early boats, the skiff is the one with the most clear-cut European roots. In Louisiana, the skiff and the pirogue were often interchangeable in terms of application. The skiff, however, can grow in size if need be, revealing that the necessary features of any folk boat, and one that we see in the bateau as well, are the ability to lengthen or widen, to become more complex or more simple, or to be made in one's own backyard or by a specialist.

Most skiffs do possess one feature that limited their lifespan in the region: they are mostly designed to be rowed, and so when it became possible to put engines in boats, most skiffs were replaced by flatboats of one kind or another, which usually had hulls capable of planing, making it possible to take greater advantage of the increased power of an engine. While they are treated in the various chronicles and catalogs as different forms, the chaland, the bateau, and the flatboat are, from one perspective, remarkably similar in nature. Apparent differences in terms of sheer or rake, or in slightly wider waistlines or bow tapers, obscure the fact that they all offer blunt bows and sterns and fairly simple sides.

The chaland, often the least interesting from a design and construction point of view, is perhaps the best example of the form. It is "crude by comparison with other boats of the region," even during the time of Knipmeyer's survey:

> The typical *chaland* is perfectly rectangular. The non-flaring sides run straight to a square bow and stern, and the end boards are vertical. There is no sheer; instead the bottoms of the ends have an abrupt, angular slant, which serves the purpose. There may be several seats or none. The extreme ends are often covered with boards about one foot wide, which are nailed to the top of the gunwales. The *chaland* is from ten to fourteen feet long, two and a half to three feet wide, and eight to twelve inches deep. A few are a little larger, and some have a small triangular external keel. (1976, 131)

The chaland is about as basic as they come, but in being basic it makes the point that sometimes what you need is a tub that floats, not a sleek shell, especially when the tub is easy, and inexpensive, to manufacture. This availability of a tub-like form will be important to the makers of the crawfish boat when the time comes: quite often what tradition does is allow, or authorize, certain forms. Without the chaland, perhaps, with only more complex hull forms available to them, it is possible that the makers might not have attempted hulls of their own. With the chaland, they had permission to go as basic as they needed to go.[46]

Beyond the chaland, there is also the flared sides, often sweeping up and in to form a bow slightly raised above the line of the rest of the boat, of the flatboat. Although it requires a bit more skill to make, its form still allows for a variety of adaptation and applications. Malcolm Comeaux argues that the flatboat's origins are probably European, since barges of this type are quite common both in Europe and, later, in the eastern United States.[47] They make their way, at some point, to the Mississippi, perhaps by passing from the Susquehanna River to the Ohio River, somewhere in Pennsylvania where the two watersheds are neighbors or perhaps by simply making their way down the Atlantic Coast and up into the Gulf of Mexico. However it happened, barges were commonly called flatboats and at least some flatboats in Louisiana were called *chalands,* French for *barge*—indeed, Comeaux notes that Albert Bourque of Bayou Benoit once told him that a chaland "was nothing more than a small barge" (1985, 175).

In addition to its reasonably straightforward construction, the flatboat enjoys the advantage of being a very stable platform on which to work and from which to fish, a factor that has kept them in production in various guises since their introduction into Louisiana. As Comeaux notes, the flatboat has almost as many names as it has forms: joe boat, john boat, mud boat, gas boat, bateau.[48] The most locally renowned version of the flatboat is the *putt-putt,* which has sides that flare out from a three-foot bottom to a little over four feet. Powering the craft, originally, was one of the first combustion engines to be light enough to be put in a boat, a one-cylinder, two-stroke engine mounted inboard that made the boat's eponymous "putt putt." Like other folk boats, the putt-putt scaled well: Greg Guirard and Ray Brassieur note that some of the boats stretched as long as twenty-four feet.[49]

It was during the First World War and the years that followed that combustion engines really began to develop the power-to-weight ratio that made it possible to put them in vessels that needed also to float. Before then, the engines that were small enough to fit in a boat were usually too weak to be of any practical use. As outboard engines improved in efficiency and power, the simplicity of buying a hull or building a simple hull of one's own and then hanging an already assembled engine off the back of it became increasingly popular in south Louisiana, and by the 1950s, shorter boats powered by outboard motors were beginning to displace the putt-putts. In addition to being a self-contained unit, outboard engines could be raised during the course of operation to clear a fouled prop and, since they hang off the back of the boat, there is also more room in the boat for people, gear, or cargo. And because the entire motor turns, they turn easily, too, with no need for a separate rudder assembly, further simplifying the overall construction of the craft—the idea that the means of propul-

A putt-putt boat participating in a parade on the Bayou Teche during the Wooden Boat Festival in Saint Martinville.

sion are also the means of directional control will prove important in later developments.

Although outboards were wildly successful in south Louisiana, as they were elsewhere, they did not necessarily work in all applications. They were, and are, especially problematic in places where land and water are mixed up, confused. Places like shallow lakes, which, during some parts of the year, amount to little more than mud flats. Places like marshes, where, if there is water, it is covered in a dense nap of marsh grasses. Places like bayous and basins thick with vegetation. In these places, the outboard engine cannot go, and residents had to seek alternatives.

Airboats were certainly one solution, and have been available since early in the twentieth century. They enjoyed some use, but they seem never to have enjoyed widespread popularity in south Louisiana for reasons that are mostly unclear. Perhaps they are seen as too dangerous, or too noisy, or too unwieldy. When asked, most people gave one version or another of a shoulder shrug, indicating that the airboat just never achieved much purchase within the network of ideas about boats. Residents of the area seem to prefer to keep their means of propelling a boat down below and not up above.

One response that emerged in the sixties and seventies was to return to the putt-putt form but to use a more powerful engine. Many a hunter got to his blinds in the marshes in a mud boat, which was built around either an air-cooled or water-cooled engine placed amidships. (Air-cooled engines were typically repurposed from used Volkswagen Beetles.) Harkening back to the chaland, the mud boat, as they were called locally, possessed a simple hull made of sheets of plywood—often marine plywood, but not necessarily; these were craft expeditiously built, and so it was not unusual for the plywood, and the paint, to be whatever was handy. Inelegant as they might have seemed to some, mud boats could power their way through amazing amounts of grass and earth: hanging one's head over the back of the boat, it would sometimes look like they were plowing, creating a small furrow as their wake.

The idea of a more powerful engine driving a propeller closer to the surface of the water stuck, and eventually, in the late seventies, as some were experimenting with a drive mechanism that consisted of a wheel that ran along the bottom of a pond to drag a boat along, someone struck upon the idea of taking the lessons learned from mud boats and transforming the outboard engine into something more useful.

One such experiment was conducted by Jimmy Boulet in Larose. Through the late fifties and early sixties, Boulet had made a decent living selling insurance to his neighbors along Bayou Lafourche, but when one year brought too many storms with too many claims, the company for which he worked decided to pull out of the region and offered him the opportunity of keeping his job, but only if he would move north of Route 190.[50] Having grown up along the bayou, Boulet was not terribly interested in leaving the life he knew and loved behind, and if that meant leaving the career he knew and liked, well, he would find something else to do. Neither trapping nor fishing were an integral part of the Boulets' family tradition, but he was lucky enough to be able to turn to crawfishing just as the commercial market was beginning to develop. Like many residents "running traps" in flooded woodlands and meadows, Boulet pulled a pirogue along to carry bait and the harvested crawfish, but, as he told his friend Lloyd Songe, "I was getting tired of pushing that pirogue around, and I'd sure liked some way of mechanizing my crawfish farming." Songe expressed an interest in helping Boulet out and asked him whether he had an aluminum hull. Boulet gave Songe the hull dimensions, which was only twelve feet, and Lloyd told him he wanted to weld something onto the back of his boat. What Lloyd did was weld a stave onto the back of the boat, upon which he balanced a small engine, a little three-horsepower one. The innovation was not in power, but in an alternate engine mount, but that was enough. Boulet could power his way up to a trap and drop the handle, which would pull the motor out of the water. The boat, which still had a little headway, would glide along gently and slowly while Boulet serviced the trap, and when he was done all he had to do was pick the motor's handle up again and power his way to the next trap. Boulet fondly recalls using this rig for as long as he crawfished.[51]

So far as the historical record suggests, Boulet and Songe's innovation did not garner widespread interest and adoption: I cannot find any photographs or mentions in various media. This particular path forward would lie fallow for a short time while innovation in the crawfish fields focused elsewhere, as we see next, but there are a few more steps taken along this line of development that deserve some mention here before we leave the larger history of boats that go on land and water in Louisiana for the smaller one of the crawfish boat.

One reason to linger a bit longer on this moment in history is to reflect on all this experimentation going on. As both Ancelet and Brassieur have noted, the ability to work in metal increased and spread significantly during this time. It seems especially the case that this growing ability developed along the southwest coast of Louisiana, which is miles deep thanks to all the bayous and canals that crisscross from Interstate 10 south, which was in the midst of developing into a hub for offshore oil exploration in the Gulf of Mexico. Thousands of men learned to cut, weld, and finish metal structures of all kinds, and they brought those abilities home with them. During this time, there were, by all accounts, as many custom handcrafted barbecue pits made from assorted metal tanks and barrels in backyards as there were store-bought grills. Although the latter might be more polished in appearance, they were nowhere near as rugged and they did not, in a culture that prized cast-iron cookware, cook as well as the heavy-gauge steel creations that emerged from shops on weekends. It was these after-hours and weekend projects that resulted in a wide variety of creativity. Certainly one outcome was the crawfish boat, but another must be what will eventually become the surface-drive engine. The man who initiated the now widely popular form in south Louisiana is Warren Coco.

Coco's great love is duck hunting, and almost everything he has done, the way he tells it, revolves around making it easier to get to where you want to hunt, improving the land on which you hunt, or allowing you to work more quickly so you can have more time to hunt. How he came upon the long-tail engine form is just such a moment; Coco noted, "The way I got into this business is . . . a friend of mine and I were both duck hunters, and he joined a hunting club near Hackberry." Membership was expensive, too expensive for Coco, but his friend really wanted him to come and enjoy the great hunting that the club was having.

And then his friend told him, "I got this funny little engine over here, and I want you to build me one. Someone was in Vietnam and came back over here and built one. He explained it to me, and I told him he was crazy," Coco remembered. He then continued: "So I went over there, met him; we went hunting; we come out and stopped at this boat shed. I got out and looked at this thing. I laughed so hard that I about fell on the ground. But you got to realize what I was looking at. It was an old, 1950s model Briggs and Stratton engine. Didn't even have a pull start. You had to wrap a rope on it to start it. The U-joint was on a 1950 Evinrude bracket, and it had a rubber hose on the bottom for a bearing. And it was using the same little propeller we used on mud boats. I had built several little mud boats. I knew I could build this, but I knew it would never work."

The mud boats he had built were small affairs—slender, fourteen-foot pirogues with squared sterns to which Coco added eight-horsepower inboard engines. Fancy was adding an aluminum tunnel over the shaft that ran from the engine and through the transom to power the propeller. The boat in front of him had a similar-sized engine, but it was mounted on an outboard frame with a long tail, much like ones seen all over southeast Asia.

Coco gathered all the parts he needed and built the first one in two months. It weighed about four hundred pounds. "It had a ten-horse Kohler engine, and I ran it. It worked so good." He refined the piece by putting on an eight-horsepower aluminum engine, which helped get the weight down. Making it lighter was critical to making it a viable product, and by focusing on getting rid of as much weight as possible, Coco quickly got very close to what would become his production model. In fact, the third one he made, he sold: "After I built the second one, I quit my job. I thought there must be a market for this. A lot of people told me I was crazy, that I would never make it. . . . But I just put my head down and never stopped. I started in my friends' parents' backyard. I quit my job and we incorporated in August 1977."

At that point, Warren Coco was only five years out of high school, having graduated and gotten off to an inauspicious start: "When I got out of high school in 1972, I couldn't buy a job. I almost joined the army, but a friend of mine's father worked at Kleinpeter's Dairy. They wouldn't hire me because I didn't have any tools, but [my father] got them to hire me because I could work out of his toolbox until I got enough money to buy my own tools." Coco's time at the dairy operation—which ran its own cows, processed its own milk, and delivered it directly to customers and to regional stores—served him well. The variety of machines that needed to be repaired and the steady stream of work kept him busy. Despite servicing seventy-two milk trucks, keeping all the packaging and bottling equipment running, and maintaining the dairy farm's equipment, Coco wanted more, and so it was not long before he started going to trade school and

Warren Coco standing in front of a large break in his Baton Rouge shop. Coco's Go-Devil motors have long dominated the surface-drive industry, an industry that he should be credited with creating.

learned skills like arc welding: "I only had a high school education, but I had something between my ears and I could work with my hands. I learned how to weld on my own."

It seems likely that if it had not been Go-Devil, the eventual name of his product and his business, it would have been something else, but it was a boat to get into and out of flooded fields and woods that was before him, and in his very first year of production, he and his partner sold 60 units, almost every one of them with the eight-horsepower engine. In the second year, he sold 160. The third year, 180. Coco remembered, "I welded every one of them myself, up to number five thousand. I welded every one of them. When number five thousand came around was when the first jig was built. I couldn't train anybody to build them the way I was building them." Within a few years, the business was doing so well that they bought land of their own on which to build a shop and left the backyard of his partner's parents, where they had initially set up shop. Eventually, Coco would buy out his partner, at practically the same time his Go-Devil engines faced their first significant competition, the Pro-Drive surface-drive engine.

Although both boat forms were roughly gestating during the seventies, the two chronologies could not, in terms of form, be more different. While there was a lot of experimentation, as we see next, around the crawfish boat drive, with Ted Habetz eventually developing the preferred solution in 1983, Warren Coco seems to have developed the preferred solution for surface drives at the very outset, and the economics were such that everyone simply purchased a Go-Devil if they needed such a drive. One difference between the two may very well be differing social contexts: there was a great deal more cross talk among the crawfish boat drive experiments, and they were working in a perhaps more restricted region in terms of geography. By comparison, the surface-drive field seems to have been more isolated: clearly there were individuals like Jimmy Boulet and whoever built the drive that was the prototype for Coco at work in garages, shops, and equipment sheds, but they seem to have had little information about what others were doing, and so they could not learn from others' mistakes and adapt/adopt others' successes.

Coco's Go-Devil was the only solution, as far as the market was concerned, for sixteen years, and then, to everyone's surprise, a pair of brothers in Loreauville suddenly presented an alternative surface-drive solution. They were Brian and Kenneth "K. P." Provost (pronounced *pro-vo*) and their outboard was called Pro-Drive. Like Coco, the Provosts were themselves avid sportsmen, and thus were in some ways experimenting in front of their primary audience, themselves. Also, like Coco, they had no real idea how many others were just like them: frustrated by the current state of shallow-water drive technologies and longing for something more powerful and more robust. None of this detracts from the genius of the original Go-Devil, but its long boom arm was simply unable to handle certain challenges.

It was getting rid of the long tail of the Go-Devil that was the key to shallow-water drives, but the idea gestated for almost a decade, sitting quite literally in the long grass behind K. P. Provost's shop. His brother Brian convinced him to haul it out of the grass and to work on it, where it remained a "slow-time" project for the next few years. Eventually the drive shaft, which was the problem, took shape and the first motor and boat premiered at the Superdome in 2003.

To say that the Pro-Drive changed the market overnight is perhaps overstating the case, but not by much, and perhaps the greatest proof of that is how many competitors attempted something similar, usually with a complex arrangement of belts—something we will see again in the development of the crawfish boat—and not with the geared axle drive that is, in fact, the subject of a patent by Pro-Drive.

Go-Devils, both long and short, and Pro-Drives—as well as Gator-Tails, Mud Buddies, and a couple of other motors—are used primarily in swamps and marshes, where the way can be seemingly blocked either by too much vegetation or too little water. They are popular among hunters and some fishermen, and there are a few crawfishermen working the Atchafalaya Basin who use them, although quite often their traps are along waterways just as easily accessed by conventional outboard motors. The shallow-drive motor and boat are thus not competitors to the crawfish boat, but companions, a parallel industry that was born and matured at roughly the same time as the crawfish boat, and probably deserving their own history. For now, this brief glimpse will have to do. It is time to jump back to the seventies to see other developments in amphibious thinking.

EMERGENCE

Dexter Guillory remembers being a young teenager growing up in Ville Platte and every summer traveling with his family to Pointe Bleu, where he and his cousins would hang out in the family's cotton gin. While the gin itself was powered by a diesel engine, the cotton came into the facility loaded on mule-driven wagons. In the fields, the bolls themselves were still plucked from the plant by human hands. With a look of wonder on his face, Guillory told me, "And that was 1960."

The wonderment, of course, comes from just how much agriculture in Louisiana, as elsewhere, had remained the same since the Middle Ages, which had witnessed the second agricultural revolution thanks to the development of the moldboard plow and the horse collar.[52] The moldboard plow, the plow most of us imagine when we imagine a plow, remained in use until the mechanization of agriculture replaced it with the disk plow. The horse collar made it possible to harness horses to heavy tasks and not choke them in the process. (Pulling a chariot around a battlefield or a coliseum is fairly easy compared to pulling a plow that wants nothing more than to wedge itself into the earth as it is pulled through a field.)

During the next millennium, refinements to these technologies occurred, but they remained essentially the same. In the early nineteenth century, John Deere refined the plow by casting it entirely out of steel, which not only made it stronger but also slicker than the iron and wood plows being used in the Midwest. Later, in the beginning of the twentieth century, the first attempts to replace horses with tractors began, and is the origin of the idea of "horsepower" as a unit of measurement and not of actual horses. However, the future did not spread uniformly across the American landscape and thus it was possible to see what was essentially a medieval form of agriculture being practiced in those parts of the country where industrialization—in the form of mechanized equipment; synthetic fertilizers, herbicides, and pesticides; and scientifically bred crops—had yet to occur. This industrialization is usually known as the third agricultural revolution.

Dexter Guillory was born at the right time and in the right place to be a witness to the transformation, and he has lived long enough not only to see but also to participate in the local agricultural revolution of crawfish being grown in fallow rice fields, extending the earning potential for farmers and expanding the production of a food source. Guillory's current observation post is at his desk at Riceland Crawfish, the company he helped to found in 1984. Two large, plate-glass windows give him simultaneously a view onto Route 190, the four-lane highway that passes through his adopted home of Eunice, and of the crawfish-processing operations that are the heart of the company.

Riceland was not the first crawfish-processing facility in Eunice, but it is the one that has endured. The first plant was called simply Crawfish Processors, and it arrived in the mid-1970s, bringing with it new economic possibilities. Prior to its arrival, if you caught crawfish out on the Louisiana prairies, you had to drive to Henderson to sell them. Perched on the edge of the Atchafalaya Basin and with ready access to Interstate 10, Henderson is a great location for basin fishermen to deliver their catch. It is, however, at least an hour's drive, if not an hour and a half or two hours, for someone from the Eunice area, which makes it not only inconvenient but also stressful on sacked crawfish, which travel best in refrigerated trucks. When Crawfish Processors opened, Guillory was among many others who actively turned their rice fields into crawfish ponds.

"Only a handful of people had ponds before that," he said.

Guillory, like many others, started crawfishing as an avocation, something to do that was both enjoyable, because it was part time and discretionary, and because it was profitable, putting extra money in his family's wallet, making it possible to take a vacation, to pay off a mortgage more quickly, or to put a few extra presents under the Christmas tree. Like others, he already had a full-time job, as marketing manager for the Winn-Dixie in nearby Ville Platte, the town in which he had grown up as the son of two professionals. He commuted the twenty miles from the home he and his wife had built on a small, hundred-acre farm that she had inherited from her grandparents. Because the farm itself had a fair amount of low-lying land, it had been crawfished before, perhaps in much the same way that he remembered crawfishing with his own father, when they would fish either for pay or on shares, getting as much as two cents a pound at the end of a day's work.

Dexter Guillory.

Guillory started small. He had to. He had to be at work in Ville Platte at eight every morning, and so he and his wife were limited by what he could run at four in the morning before he left for work, and what she could run in the afternoon, often with their children riding in the boat with her. The money was good enough, though, that it wasn't long before he and a few friends decided to open up their own buying station, acquiring crawfish from the Eunice area and selling them to someone in Pierre Part who had established clients and distribution channels.

By 1980, he had expanded his home operation to use all of the farm's one hundred acres and was making as much money fishing as he was working at the grocery store. It was then that he decided to work for himself. As he built up more experience, learning the hard way that rotating crops was also good for crawfish ponds because it kept aquatic grasses and water-tolerant trees in check, Guillory also seems to have realized that controlling his own destiny also meant controlling more of the process. It was that insight that led him to found Riceland Crawfish in 1984, which boiled, peeled, and bagged the crawfish tail meat and employed fifty or so employees seasonally from the very beginning. Looking back, Guillory observed the parallels between his own development as an entrepreneur and what everyone else was doing and noted, "The whole industry developed in ten years' time."

As a longtime observer of the local scene, Guillory remembered a time when crawfish were caught pretty much the way they had always been, using drop nets in low-lying areas that flooded. As farmers began to realize the potential profit in trapping crawfish in fields that were otherwise not getting used, they began to use the same methodology, placing drop nets in a field much as one would in a wooded area, walking around in waders to place the nets and then later to empty them of crawfish.

"We would all use drop nets. We would put out fifty to a hundred and just keep running the circle," Guillory said. As the practice grew, with more farmers trying out more land each year, individuals began to seek out ways to make the process more efficient—to catch more crawfish. The problem, of course, is that drop nets are not traps. While fine for the weekend fisherman, who is probably only looking to fill a pot or two for a family crawfish boil, drop nets are literally nets dropped onto the bottom of a shallow water pond. The nets are typically a foot and a half square, made of cotton netting, and held open by two strands of stiff wire that are bent to form two pairs of legs. This arrangement makes the nets easy to fold open for use and to close for storage, but it means that all they present to the crawfish is about two square feet of netting lying on the pond floor. The nets are baited, usually with scraps of chicken or fish, and left to stand for a time. Gathering up the nets requires both stealth and speed: one must approach the net smoothly and slowly and then grab it up quickly once it is in reach so as to give the crawfish as little time as possible to scoot backward off the net, something they can do amazingly well.

What was needed was a trap that made it impossible for crawfish of a reasonable size to escape. One such trap was readily available and already in use in the Atchafalaya Basin, the historical home for crawfishing in Louisiana. The pillow trap earns its name from it being about the same size and shape as, well, a bed pillow—though, to be fair, it is probably better to compare it to a king-sized bed pillow. The trap is relatively easy to make and consists of a sheet of metal mesh, usually nylon-coated now, that has been rolled into a cylinder. The cylinder is then flattened to form an ovoid. The bottom is closed up, except for the two corners, which are pushed in, often with a beer bottle, to form a funnel into which crawfish can crawl. The top of the trap is either folder over or held closed—sometimes with a clothespin or two. In the basin, with deeper water, the trap is usually tied to a tree. If there is a current, then the trap can be completely submerged, and the line acts both as a visible marker as well as a convenient way to retrieve the trap. If the water is still, then a portion of the trap needs to be kept above the water in order to keep the crawfish from suffocating.[53]

Using the pillow trap in fields, farmers gathered their catch at first by pulling a large tub behind them, and then later by buying lightweight aluminum john boats and pulling them by handles mounted to their bows.

"You could walk with a boat and not fall in the water. You put your bait in the front and either emptied your catch directly into the boat, or I preferred to empty into five-gallon buckets, which made it easier to sack up the crawfish later." Using the buckets as his guide, Guillory knew that for every two buckets, he had a sack of crawfish.

Obviously, with no trees in rice fields, and with most fields being too deep simply to lay pillow traps down, some way had to be found to keep them upright. That meant staking and then either tying the trap to the stake or carefully passing a sufficiently thin, and also sufficiently strong, stake through the trap. The result was reasonably satisfactory, and it worked well as long as you were dragging a tub or a small boat behind you as you walked through the fields.

The story of walking a flooded field to collect a day's catch is a common one among older crawfishermen from the area. Some remember their first forays of walking with a five-gallon bucket in each hand, one to collect the catch and one to rebait the trap. Others remember pulling washtubs or plastic toddler splash pools that had been pressed into service. Almost everyone eventually graduated to a small aluminum or fiberglass skiff, usually pulled from the front. One extension agent from this era, Dwight Landreneau, has a photograph of a one-horsepower skiff; that is, a skiff being pulled by a horse.[54] As the crawfish were dumped into the boats, the boats got heavy. And the weather out on the prairies, when you are trudging through either knee-deep or thigh-deep water, is almost always too cold or too hot. Surely it would not only be nicer, but also more efficient, to ride with the bait and catch rather than pull it through the fields, where the boat would knock against you or pull from you, depending on which way the wind was blowing.

"The traps and the crawfish boats grew together," Guillory reminded me.

With the switch to pillow traps, the traps had become more efficient than the collection process. As catches increased, so did the possibility for getting through the day's catch more quickly. Tubs had to be emptied regularly. Five-gallon buckets of bait had to be refilled regularly. Stopping to unload or load buckets or basins increased the time spent in each field, lessening the number of fields that could be run each day. As farmers caught more, they made more, and they made more in a crop that was much more straightforward than the complexities of planting rice or soybeans, which often require loans, insurance, authorizations, and a variety of other paperwork that most farmers do not relish. It is not why they farm; they farm because they enjoy the simple equation of putting work into something and then getting results out of their work. They are also comfortable with risk: sometimes all your work, no matter how long and how hard and well planned, can be wiped out by a hurricane or a drought. Catching crawfish was like returning to a previous age in agriculture, when work and nature were the two dominant factors and not a commodity price index that really measured the greed and/or anxiety of men in suits who worked in cities far from the hot sun and high humidity of the prairies.

Stories about buckets and tubs are so often told, I am convinced, because this was a critical moment for the aquacultural revolution that was taking place among people who considered themselves then, as most do now, agriculturalists. (This played out in how they thought about the boats, as is described later.) The image of a man towing behind him a bright-blue children's pool is funny, to the tellers most especially. They are gently making fun of themselves; at the same time, the ludicrous nature of the image also emphasizes exactly how hard the work was, how much of it there was to do, and how crippling the available tools were felt to be.

As yields increased, farmers began to want to add more power to the process. These were men, after all, who were comfortable with a wide variety of farm gear, and, just as important, most had some knowledge of and experience in making or modifying an implement in their own shops or equipment sheds.

What happened next is not entirely clear. Memories in the present must reach back forty years and try to piece together, through reference to other events—often to when a child was born or to what grade they were in—that otherwise blend together as part of "life on a farm." There are few photographs available, even from the men who would become prominent as boat makers. Few farmers take pictures of their tractors, nor do they date when they first used them. The same is true for the crawfish boats.

The first boats that were built were mechanically driven, with a small engine, five horsepower by most accounts. What exactly they were driving is not immediately clear. A number of folks have described a variety of pulley-and-cable systems that seem better suited to steerage than to powering a boat. More than once I have heard them labeled *contraptions,* as if that was all there was to say on the subject. Whatever the form of those first few protoboats, they seem to have had the drive unit in the back, similar to the modern crawfish boat, and somehow managed to make their way through a field, despite the great deal of slack in their workings—and the consensus is that they didn't work that well.

Out of this initial period of experimentation emerged, in the very late seventies, the "tiller-foot" boat. Like its predecessor, it was built on lightweight boat hulls that were commercially available with the addition of a most extraordinary assemblage: the lower part of a garden rotary tiller driven by a five-horsepower engine separated by a long boom, made in place. At first one steered the machines with a tiller attached to the assembled drive unit, much like you would the outboard motor of a boat, but eventually someone adapted the power-steering cylinder of a car so that steering could be handled more remotely.

"Greg Frugé was into race cars, and he was always fooling with that kind of stuff," Guillory recalled. "He was probably the one who came up with that and everyone just copied him."

The tiller-foot boats were used for a number of years—Guillory estimated seven to eight years but others have suggested a shorter period—but the fact that the drive units were assembled from such disparate parts meant the units were short-lived: the transmission gears, often made of brass, were not intended for such intensive use and would typically fail within a year's time. The widespread availability of the machines from which the parts came, garden rotary tillers, and their relative cheapness, a hundred dollars or so, kept tiller-foot boats in use for a while. It is also possible, it should be noted, that another possible dimension of the boat's longevity, despite its mechanical failings, was that many of the tillers came with warranties, and it was not unknown for a failed transmission to be put back into its original box and returned for replacement. In this case, it is possible that one of the factors that contributed to the demise of the tiller-foot boat was that local store owners and equipment dealers eventually caught on to the practice and began to refuse to replace tillers that had exceeded their intended use.

Just as the improved traps had pushed development of a boat that could handle their increased efficiency and capacity, the boats now began to demand traps that could keep up with the speed with which they made their way through a field.

"When we got away from pulling the boat to a mechanized boat, we moved to an open trap. You just couldn't move fast enough with the pillow traps," Guillory said. And so farmers began to modify the pillow traps: "We started by making the traps look like a trash can. We made a circle out of expanded metal, put mesh on the bottom, sewed in two funnels, and left the top completely open. You would dump it just like a trash can." Expanded metal, unlike wire mesh, is made from a single sheet of steel that first has a series of slits made into it and then it is stretched. It is often used in grates and outdoor furniture and is extremely tough stuff, making it both durable but also more difficult to work. The trash-can trap seems to have worked well enough that at least a few farmers remember using

A pyramid crawfish trap showing its three funnels: note the trap in the background staked in a flooded field.

it for a number of years, but it would eventually be replaced by the design that is now the standard, the pyramid trap.

Designed for use in the long, wide, shallow, and flat bottoms of rice fields flooded with water, the pyramid trap is a mesh tetrahedron with a cylinder on top of it.[55] Some stand about two feet high, and some are made a little taller: there are traps with extended chimneys that stand four-and-a-half feet high. Each of the three bottom corners of the tetrahedron, or pyramid, has a funnel in it, allowing more crawfish to enter, and from more directions, netting a greater number of crawfish per bait.

The pyramid portion of the trap, like the pillow trap, begins as a rectangle of wire mesh two feet by four-and-a-half feet or so. The rectangle is rolled into a cylinder, and a funnel is begun at one end of the seam. The cylinder is then flattened with the seam facing upward, and while this end of the cylinder is getting closed up, the other two funnels are formed at what are now corners. As this happens, the cylinder is transformed into a tetrahedron, with the mesh getting pushed and tucked as the trap gets worked. Eventually what results is the characteristic four-sided pyramid shape, with each corner open: three of the corners have the mesh pushed in, and the fourth corner has the mesh opening outward. A long-necked glass bottle, or a wooden spindle built for the purpose, helps to form the funnels more carefully, making sure they are large enough to admit the desired size of crawfish and angled upward forty-five degrees or more to make sure the crawfish cannot get out again. The most important side of the pyramid is the bottom, since in some instances a good trap will sit upright all by itself, with no need for a rod or stake to keep it upright. Staking a trap not only adds expense, but also a fraction of a second of time and energy that can, over the course of a day with hundreds of traps to pick up and place back down again, make things more tiring than they need to be.

Atop the trap is the chimney, which, as aquacultural specialists Mark Shirley and C. Greg Lutz note, is an especially useful innovation, since it acts as a "combination collar and handle on the top of the trap, which prevents crawfish from climbing out while making the trap easier to grasp, lift and empty quickly" (2009, 2). The first cylinders were made out of six-inch PVC pipe, as many still are. Although the first few chimneys were plain, and there are still a few in use, it did not take long for operators to realize that making them easier to grab from a moving boat would make the work go a lot more quickly, and someone discovered how easy it is to heat PVC enough to push in one side of the top of the cylinder to form a lip about one inch deep, just enough to catch it with fingers. This design has been solidified in the cast-plastic tops that are now widely used and are sold separately for those who want to make their own traps or for those who need to replace a failed top.

The pyramid trap, as its description reveals, is not an easy trap to make. As both the traps and boats evolved, they became simultaneously not only more efficient but also more complex. They were less available to ordinary farmers with limited time to spend on building machinery for what often was a side crop. Gone were the days when, as Guillory noted, "three or four guys would get together and make a bunch of boats. Everyone would have one, and they would sell two or three. Everybody didn't make a boat, but it was a pretty small circle of who ended up with a boat you made." Guillory also lamented that, in general, there are fewer farmers. He remembered, and prefers, a time when there were a lot more small farmers, some of them only fishing fifteen- to twenty-acre fields. There was a lot of experimentation, some of it a result of smaller operators not having the desire, or the ability, to afford a commercially made boat. The result was a lot of boats that resembled other boats in many ways, but each also bore its maker's individual mark. Some boats, no doubt, manifested real insights, some of which would get taken up by others. Some boats may have been less "inspirational" and more the result of someone managing to cobble together something that worked well enough to keep it running from year to year.

It was into this milieu that what would become the first commercial crawfish boat emerged. Its maker thought of himself more as one of the small operators, but he would soon find himself building boats on a scale that, even now, makes him laugh. Ted Habetz had no idea that he was about to drive a boat into history.

FLUORESCENCE

Louis Cramer and Harold Benoit are the quintessential complementary pair. Cramer is a tall, lean man from the German community. Benoit is a stocky, barrel-chested Cajun. Cramer speaks softly and slowly, each word measured against what may or may not be most appropriate. Benoit speaks quickly, excitedly, with a passion that cannot be contained. Together the two men formed a team that would launch a veritable revolution in crawfishing, though they never imagined it as anything more than the next logical step. Thing is, they saw steps.

Cramer and Benoit began their efforts by focusing on production efficiencies but by the time they were done they had participated in or created any number of organizations that sought to build and maintain a viable market for crawfish. They sought to turn what was then a variable windfall into a sustainable business for an area that faced an always-competitive agricultural marketplace. Both were systematic in their approach to things. Cramer was the kind of man who not only taught a subject but knew its institutional history as well. He had observed the change in vocational agriculture programs from production to agribusiness, and he wanted to apply the same logic to crawfishing. Benoit was a retired military man who had experienced the Cold War up close and understood that small things led to large outcomes. Together the two understood that although fishing crawfish out of fields occupied a nice niche for many farmers, bringing in extra cash that could offset a bad rice season, farming those same crawfish could be a viable business for others, perhaps one more stable and less subject to international pressures than agricultural commodities. They could not have, in that moment, anticipated the oil crash that would befall Louisiana's economy in the eighties nor the influx of cheap Chinese crawfish in the nineties that would drive a number of folks out of business, but it may have been their ability to put crawfish on a business footing that allowed those who did survive the economic turmoil of those two decades the chance to do so.

Louis Cramer.

Cramer started crawfishing in 1974 when he stocked a small, eighteen-acre field near his house. He started harvesting the field the following year, pulling a small, plastic children's swimming pool behind him as he went. He had seen some folks trying to use boats equipped with Go-Devils, but he felt they made too much of a mess. In 1980 he had helped to start a trade show of sorts: it began as a crawfish tasting at a dance hall between the town of Scott and the city of Lafayette. Increasing the market for crawfish had gotten him thinking about increasing production, and so in the fall of 1982 he held the first crawfish field day, a homegrown version of the larger field days held by the state university's agricultural research station.[56] He invited folks to come out and his plan was to demonstrate a crawfish buggy made in Texas; he was pretty sure it would get people talking.

Sure, everyone talked when Amos Roy of Beaumont demonstrated his machine. What wasn't there to say about something that looked a bit like the lunar rover set down in a muddy Louisiana rice field? The operator sat in the middle of four large wheels that tracked through the shallow water reasonably well. He drove right up to a crawfish trap, grabbed it and emptied it to one side. It was an amazing machine, but it appears to have been eclipsed that day by a machine that turned up at the last moment, a john boat–come-lately that was built by Tedmon Habetz, who wasn't entirely sure what he had just gotten himself into.

When Benoit saw Habetz's boat, he turned to his friend Lawrence Adams and simply said, "That's my boat." What he meant by that is a tale of two men working simultaneously fifty miles, and years of experience, apart. Both had roots in agriculture; both had professional degrees; and both had had experiences beyond their education and beyond their agricultural roots. As divergent as their educations and experiences were, their convergence on the same solution at the same moment in time suggests that there was a certain inevitability to the idea, that it was, as Kevin Kelley has observed, "what the technology wanted." The inevitability of such convergences is something for a larger conversation, but the role of hydraulics, on a hydraulic landscape, is discussed later.

~~~~~~~

It was Habetz's boat that appeared before a curious public, and so it is Habetz who is credited with inventing the modern crawfish boat. When asked, Ted Habetz says it all began in 1964, when his father decided not to drain one of the fields that had been flooded by Hurricane Hilda, and that the presence of that field in his life—indeed, his place on the Louisiana landscape—was a near miss that, fortunately, didn't occur. He was almost born in Mississippi. His father Joe and his mother Rita were on their way to buy a farm there when a real-estate agent remembered at the very last moment that the old mill area from Mariah Plantation near Loreauville was for sale. Loreauville is along the Bayou Teche, a sugar-growing alley of Louisiana, and the land that Joe and Rita bought was already planted with cane. But Joe Habetz was a rice farmer—the Habetz family is from the prairies in and around Roberts Cove. So Joe saw the crop out, but in the years that followed he built what he knew, a rice farm, digging canals to flood the fields and using the original pump from the old mill to move water.

Ted Habetz was born in Loreauville, grew up in Loreauville, and lives there now with his own family. It is a place with an amazing amount of innovation and industry in it. There are three large boatyards in the town that all specialize in building large craft that see service all over the world, in no small part because a lot of the designs were developed for the petroleum industry, which has taken the boats wherever it goes. Two of the shallow-water drive manufacturers, Pro-Drive and Gator-Tail, call Loreauville home, and Ted Habetz is in fact friends with K. P. Provost of Pro-Drive. Their families have lunch at the same restaurant on Loreauville's main street every Sunday after church.

Possessed of a tall, wiry frame, Habetz has a penetrating gaze that emphasizes that his curiosity drives past daydreams into application. In middle school he attached a motor to his bicycle. In high school, he spent several years working on a rocket that he eventually entered in the science fair. To this day, he remains involved in science-fair projects and proudly points out the display area he set up in his office's window to show off the genius of the students with whom he works. Habetz loves solving problems and is good at it. The engineering firm, which he helped found, has an impressive staff focused on a wide array of projects, all of them complex and in need of a careful eye.

Habetz's attention first turned to crawfish in 1964, when he was twelve years old. It was the year that Hurricane Hilda struck south Louisiana and flooded one of his father's rice fields, the same as it had done for Jimmy Boulet across the Atchafalaya Basin. "When the hurricane came, the whole back of the field just flooded," Habetz reported. "So my father left it like that. And that's how we had our first crawfish pond. It was because one of the hurricanes that came. I was twelve years old, and I used to walk the ponds. We always had crawfish after that." Habetz helped after school and on weekends with the crawfish, which was mostly a pastime for his father, a way to have crawfish to eat when he or his friends wanted some. When his schooling was complete, he decided to get an engineering degree at Louisiana State University, which he described as the first four years of his education. The next four years, he noted, came at the hands of the men who ran the material lines at his first job out of school, at the Pittsburgh Paints (PPG) plant outside Lake Charles. It was there on the factory floor, with union tradesmen as his guide, that Ted received his education in hydraulics, gearing, and distributing power effectively and efficiently.

"[After] I went to school at LSU," Habetz recalled, "I went to work at PPG. One of the units I was assigned to was the silica pigments plant over there. Pigment. Paint. The stuff that stays on the wall, that's not the color. They put the color on something that sticks to the wall. Silica pigments made out of sand, glass. You mix sand with caustic. It goes into a screw conveyor, and they put it into a big oven. They cook it. It's melted for twenty-two hours by the time it gets to the other side of the oven. Huge. And that stuff is coming out in a big stream. Looks like molten metal. And it hits this trough, this conveyor trough that has water in it. And it's rumbling. They've got this molten glass hitting that water. It shatters. It shatters so much it dissolves in the water. Water glass. Then we thickened it up, dried it out, stuff like that.

"But on the front end," Habetz continued, "where we had to blow the air into the sand with the caustic and the fire, that was very, very, very tough. We used to turn the screws—we had three different screws that we were feeding into that—with hydraulic motors. Terrible working conditions. It was hot. Dust tore up the seals. I did a lot of maintenance work back there. I learned hydraulics."

After ten years at the plant, and some twenty years after Hurricane Hilda first flooded his father's field and turned it into a crawfish pond, Habetz found himself back home, working at a salt mine (quite literally): "I moved back over here and started working at the salt mine, still doing engineering. Oh, man, I'd get off at three o'clock. Plenty of time. So I started crawfishing over here in my dad's fields." Habetz smiled and recalled the particular arrangement he had with his father: "The first cut, which was

Ted Habetz.

about four acres, was for him. I couldn't fish that first cut. That was his. I had from there as far as you wanted to go, as many acres as you wanted, but the first cut was his. I put two traps in that. That's it. So whenever he wanted crawfish, he'd go pick up those two traps. And that was a whole sack of crawfish. I always kept the traps baited for him and everything."

The rest of the ponds were Habetz's to crawfish, which he did, on foot. Every afternoon he would walk the ponds, pulling up traps and rebaiting them as he went. On average, he would catch enough crawfish to fill up a few sacks, which he would put in a cooler for his dad to sell. For Habetz, it was almost an equation to be solved: "Here I am walking this stuff and Bruno's got a boat. I am an engineer. So I got with a friend who was a welder and we built a boat in my old man's shop. We built it in a weekend. We had all the parts. I lived on a farm: if you need something, you can find it. We built it all out of scrap parts."

Habetz's brother Bruno had built an eighteen-foot-long, double-hulled monster of a boat that was pulled through the water by a spoked wheel turned by a worm drive pulled from a combine. Seeing the thing, Habetz thought he could do better and he and his friend, David Louviere, used a variety of angle iron and constructed something that looked, according to Habetz's memory, more like an oil rig than anything else. They had a hydraulic motor, but they hadn't yet discovered how to achieve a gear reduction using only the hydraulic system, and so they connected the motor to the wheel via a chain.

"That first one we built," he said, "I remember, had a chain and everything. Well, we had a problem with that, so, you know, I got a bearing on one side. So we're going to put that motor driving direct to turn the wheel, one to one, get the right motor, get the right-size pump in the back. And we're going to get our gear reduction: put a little-bitty puny pump with a big motor in the front and that's your gear reduction to get the right speed. It's done. There's no bearing, solid coupling. We made that look good. Forward, reverse."

As Habetz recalled each detail, his hands moved in the air, assembling the parts from memory, enjoying again fitting together the puzzle of pieces, but he was quick to acknowledge that the boat was a product of a partnership of ideas in the head and hands on metal: "David Louviere, he was a master at working iron. He was good. I might have had it on paper. He made it look good when we built it."

The boat itself was a partnership of components drawn from a wide array of other devices: the drive wheel was off an old plow and made of iron. The hydraulic system was based on a power-steering pump pulled from an old combine. The pump ran off a pulley, and they found a V-belt to turn it. They pulled all of this together in a weekend, and, to their surprise and satisfaction, "the son of a bitch worked." Looking back, Habetz noted that everything really came together when he stumbled across a particular valve, whose part number, SPFWZ4, he still remembered. The valve allows an operator not only the ability to put the boat in forward and reverse, but also to vary the speed of the motor—remember, the pump operates at a steady speed—and to put the whole system in "idle." His stumbling across it was really a function of his work in hydraulics at his day job: he had a lot of catalogs on hand and would regularly look through them, not only seeking parts for particular projects but also to learn what was available. In this instance, his curiosity paid off. The valve was so central to their design that for the first few years, Habetz and Louviere ground off all identifying part numbers. (That worked for about three years.)

Shortly after that, Louis Cramer showed up and invited Habetz to show his boat at the field day he and Harold Benoit were organizing. Habetz was interested in going because he had heard about a new style of trap that Amos Roy would be bringing with him. Habetz was, at the time, using traps based on his father's design, which had four inlets and a bait cylinder, all made out of plastic mesh. Although there had already been some interest in Habetz and Louviere's boat before the first crawfish field day, that day solidified their hold on the larger imagination. The two men brought drawings to make it clear what they were up to, and that cemented the idea for many present.

Habetz ended up taking orders for four boats that first year, selling most of them to J. F. Noel of Kaplan. Later he would form his own company, Crawfish Combines, which would end up building three hundred boats during the next ten years. His partner in the business was Louviere. Their business was housed in Ury Louviere's Welding Shop, which belonged to David's father. At night and on weekends they transformed first "green hulls" from Ouachita and later their own custom-built double-raked hulls into "crawfish combines."

Habetz emphasized that their focus at first was only on building "the rig" that fit inside commercially available hulls: "We built a frame on the inside of the boat, because we were pulling. We would formfit that, so you were always pulling on the inside of the boat. We would drill that. And we built the power pack. That's how we got started with our power pack: we made them to fit those Ouachita hulls, so we could just drop that down in there." They ran the hoses on the opposite side of the boat from where most people worked: the dominance of right-handedness put most operators on the right side of the boat, so the hoses ran up the left side from the engine to the wheel in the front. The engines they eventually decided on were made by Honda. Engines by Briggs and Stratton and by Tecumseh were available, and were usually cheaper, but everyone agreed that the Hondas ran better in the cold. Crawfishing typically begins in late winter, and so an engine that started and ran reliably in the cold was preferable. The horsepower on the engines started at eight with the Briggs and Stratton engines, but moved up to ten with the Honda engines.

In addition to a change in engine manufacturer, Habetz and Louviere eventually found themselves confronting the production of hulls: they would begin to build their various assemblies prior to the boatbuilding season, which is in the fall, ahead of the crawfishing season, which lasts from late winter to late spring. They discovered, the hard way, that retailers tended to draw down any remaining inventory they had over the same period of time, usually over the course of the summer, with the understanding that anybody who was going to get a new boat for the year would already have gotten one. That meant that the pair's production cycle did not match well with the local retail cycle, and they found themselves with drive assemblies and no hulls on which to put them.

"We weren't interested in building a boat," Habetz said. "We were interested in the rig and putting it on a regular old aluminum riveted boat." Nevertheless, it wasn't long before they found themselves in the boatbuilding business. "That's when we started getting into the welded aluminum boat. We'd buy an eight-by-twenty sheet of aluminum. We'd cut it fourteen foot, so we had a six-foot drop. Then we'd have the whole hull just crimped, so we'd just pull up and put a rake on each end. So the boat would look the same forward and backward, because sometimes you'd back up as fast as you were going forward." How the pair pulled all this off while maintaining full-time jobs elsewhere is something to admire.

"We worked Monday night, Tuesday night, Wednesday night, and

Thursday night," Habetz said. "We'd take off Fridays. And then we'd work four hours on Saturday. And then we'd make deliveries on Saturday afternoon."

Some of the deliveries were quite far away, including at least one to South Carolina. For these deliveries, they hired Louviere's younger brother to drive. They went on a lot of the deliveries closer to home not only because they wanted to meet the people buying their boats, but also because often they needed to show them how to operate the boat and how to load and unload it from a trailer. A lot of these purchases, Habetz told me, were over the telephone. People were buying boats, in some cases, sight unseen: they had simply heard about the magic and they wanted to possess it for themselves.

The boats they built came to be called "pull boats" or "front-wheel-drive" boats, in large part because it was the kind of boat that worked well for Habetz himself, who, like a lot of farmers on the eastern side of the prairies and the western edge of the Atchafalaya Basin, had a few, large ponds. In Habetz's case, he had two, which meant he only needed to cross a levee once per run: he would simply alternate which pond he left the boat in and use it as the starting point for the next day. Minimizing the need to cross levees meant that he could make do with hauling his boat over a levee using nothing more than a hydraulic-powered winch.

"We had all this hydraulic power, we just put winches on the front," he said. "We'd pull to a mobile-home anchor." The anchors are made in much the same way as grain augurs: a piece of helical flighting wraps around a steel rod, turning it into a giant screw that can not only be fairly easily turned into the ground but also "unscrewed" when it's time to plow a field for rice. There is no permanent structure to worry about, and there are no postholes to be dug or, later, to be filled in.

Habetz and Louviere continued to build boats, part time, for ten years, finally stopping sometime in 1993 or 1994. In that time they made close to three hundred boats, some of which traveled to Central America to be used in shrimp farming. What pushed them out of the business was a powerful combination: they had saturated their own local market for boats and Chinese crawfish were beginning to saturate the American market. Prior to that, crawfish trappers and farmers made a good living, but starting in the early nineties, the price pressures brought by the influx of Chinese crawfish made it difficult for the small operators, especially those focused solely on crawfish, to stay in business. As business slowed, the two men decided they had had enough fun; they were ready to do something else.

Ted Habetz remembered talking with Harold Benoit at that first field day, and while both men had struck upon the idea of the hydraulic boat at about the same time, they both also had designs that were influenced by their regions and their customers' working conditions. Habetz and Louviere's boat had a long arm at the end of which hung the wheel; Benoit's boat had a shorter drive arm, which Habetz likened to a "bicycle wheel." In calling it that, he was trying to describe the limited traverse of Benoit's wheel compared to his own. The longer reach of Habetz's drive unit meant the wheel itself could drop to a greater depth and still be in contact with the bottom of a pond.

"A lot of places we were selling boats to were using cleared woodland," he said. "They had big slews and gullies, and sometimes they'd have to cross, going from one field to another, a huge canal in the middle that they had flooded." Instead of flooding individual cuts within a rice field, these operations would flood to the big levee that runs around the perimeter of the entire field. "They were fishing in the deep water, too. Some of them had six-foot-long traps, rather than the thirty-six-inch traps. We had a different clientele over here as opposed to rice farmers over there."

Harold Benoit grew up in Thornwell, and he remembered catching crawfish when they drained a rice field at the end of a season, placing a sack across the spot where they broke the field levee. Benoit credited his father-in-law, Alphé Simon, with his introduction to crawfishing as something more than a casual pursuit. Simon and his wife, Lily, had a farm just outside Morse, and he had started harvesting crawfish in his rice fields in 1959. Out of almost four hundred acres, Simon set aside thirty acres for crawfish. The fields on those acres produced some of the biggest crawfish Harold Benoit had ever seen, some of them approaching five to six crawfish per pound. More than anything, Simon loved the outdoors. He loved to go crabbing and fishing and frogging. For crab and fish, he headed south to the Rockefeller National Wildlife Refuge, where he fished along Hackberry Ridge. For frogs, however, he needed to look no further than the intricate network of canals that laced through the area at the

Harold Benoit.

time. If you combined his keen fishing abilities with his knowledge of the local landscape as well as his adventurous personality, you had a recipe for someone likely to find a way to grow crawfish in his own fields. It didn't hurt that having crawfish brought people to his door, and he liked visiting with people.

For bait, Simon used fish caught from the canals. He placed traps along the levees that form the smaller sections of rice fields, the cuts. To collect the crawfish, he walked along the levees with a bucket and emptied the traps as he went. Fishing from the edge of the cuts did not net a lot of crawfish, but it was enough to provide crawfish for family and friends and to sell to the occasional acquaintance who asked. The methods and technology did not change much over the next decade and a half, when his grandson Kevin began to crawfish in the mid-1970s, but when Kevin's father and Simon's son-in-law, Harold Benoit, stepped in to take up where Kevin had left off after the latter had graduated, things began to change. Of course, the same thing that made the traps easy for humans to reach also made them easily reached by other animals, like raccoons and rats. Harold Benoit grew up in Thornwell, about five miles due west of Lake Arthur along an old railroad line, and his childhood and adolescent memories of "crawfishing" was of catching crawfish when they drained a rice field at the end of the season. His family would scoop them up with a sack placed near where they had broken a field levee or, even more conveniently, placed over a drain pipe. His wife Juanita remembers much the same sort of thing happening on her family's farm outside Morse.

And that was pretty much how crawfishing was done when the two met, fell in love, and married in the 1950s, shortly before Harold left Louisiana to serve in the Air Force. Juanita stayed behind, living with her family, and she remembers quite clearly the meal her family served her before she left to join Harold in Canada, where he was stationed: it was a heaping plate of crawfish étouffée made with crawfish caught in that thirty-acre pond of her father's. Over the next twenty years, Juanita returned time and again, whenever Harold was stationed somewhere she could not go, and she watched her father's business grow and grow. By the late seventies, she remembers, the demand was so great, they were taking orders three weeks in advance.

This timeline places Alphé Simon as an early entrant into the commercial crawfishing scene. Starting in 1959, Simon had found a way to combine his work as a rice farmer with his exploits as an outdoorsman, along with his gregarious social nature, by inviting the public to go into his rice fields, after he had harvested and reflooded them, to catch crawfish for themselves using their own traps.

In the mid-1970s, Harold and Juanita's son Kevin began to work his grandfather's field while he was still in high school as a way to make some money of his own. He fished the fields the same way everyone else was doing it, by placing traps around the edges of the field where he could get to them easily. He walked along the shoulder of a road or the crest of a field levee and emptied the traps first into a five-gallon bucket he carried in his hand and later into a small john boat that he pulled along as he walked.

Kevin graduated from high school in 1979 and gave up crawfishing when he took a full-time job. His father Harold retired from the Air Force in 1978 and, although he had started work as an assistant manager at a local feed store, decided to take Kevin's place in what was a decent, if labor-intensive, side business for the family. In the fall of 1979 and the spring of 1980, Harold Benoit followed, quite literally, in his son's footsteps, pulling a boat through his father-in-law's flooded rice fields, emptying traps placed along the edge of the field that had sometimes been picked over by other crawfish predators: birds, nutria rats, people.

Benoit ran traps like everyone else that year and the following year, but he spent the summer of 1981 in the time-tested Louisiana rice culture's belief that there had to be a better way, thinking about what kind of boat he could build that would not only pull him around, instead of him pulling it, but also allow him to fish the middle of the cuts. The middle had two things going for it: there had to be more crawfish there and it would be harder for lazier predators, especially people, to get access to the traps. As he daydreamed, he began to collect potential parts, scrounging reel motors from combines as well as various valves and controls.

On his very first boat, Benoit used a valve from a reel and a valve from a grain elevator off an old John Deere combine. Those valves and a three-way valve were all mounted to the floor and operated by long handles so Benoit could operate the boat while standing up. There was, Benoit points out, nothing commercial on that first boat. No store-bought parts. He built the boat over the course of the summer and fall of 1981 and waited for his chance to try it out in the fields once they were flooded. While he waited, he ran it repeatedly in his carport, testing his lines to make sure

Harold Benoit's first crawfish boat used the curved arm of a plow to carry the wheel.
Photo courtesy of the Benoit family.

Later boats by Harold Benoit used a simple piece of heavy steel plate cut into a triangle that could easily be mounted to the front of a boat. Note the use of commercial boat hulls. Photo courtesy of the Benoit family.

their connections would hold. Everything seemed to work, but when he put it in the water, one thing became immediately obvious: the boat went too fast. Benoit had no way to vary the speed, to slow the boat down. He noted, "When you gave it fluid, boy, it took off."

A year passed and then, in the fall of 1982, Benoit went to the first crawfish field day organized by his friend and fellow vocational agriculture teacher, Louis Cramer. On the day of the demonstration by Amos Roy of his crawfish buggy, Benoit remembers seeing what he called "the first combine that anybody had ever seen." He was looking at Ted Habetz's boat. Admiring what Habetz had done, Benoit recognized that he and Habetz had been working along similar lines, but that Habetz, more experienced in hydraulics, had worked out the gear ratios to make it work.

No one at that first crawfish field day on that crisp October day in 1982 had any idea that the region's entire economy was about to be changed; they only knew that someone had a working solution to a long-standing problem. But a reliably mechanized boat would significantly change the scale of everything they were doing.

By the following fall, Benoit had a working boat in the field. It was built with new parts for the hydraulic system, and he fabricated the additional steel structure he needed: the leftover plow that had been the original arm holding the wheel out in front of the boat was replaced with a triangular tongue of heavy sheet steel that was hinged at the bow. All the important pieces of the contemporary boat appear to have been in place: a high-RPM, small-bore engine drove a hydraulic pump that delivered, simultaneously, power to a ram that turned the boat and power to a hydraulic motor that moved it.

Most important, Benoit had a working boat in the field. It wasn't long before he had cars and trucks lined up along the roads that bordered the fields he worked to see what was either a crazy contraption or a crawfish combine, depending upon whom you asked. Farmers showed up in carloads to watch Benoit run his traps. As they watched, they counted traps, pounds, and hours, and the dollar signs piled up like the sacks of crawfish on the bow of Benoit's boat. When he came out of the field, farmers would tell him, "I gotta have that boat."

Benoit declined, "No, I'll tell you how to make one."

After one of these first days of spectating, one farmer came back. Bill Krielow told Benoit: "I want three, and I'll write you a check before you buy the first part." Benoit replied that there was no schedule. Krielow's response was that Benoit could make them when he had the time.

"Okay," Benoit said. "I'll build you one and see how it goes."

By the time he had built Krielow's first boat, another farmer, Larry Lyons, wanted one, and so did somebody else down in Gueydan. Looking back, Benoit commented, "And it just snowballed." In the end, Lawrence Adams got the first boat Benoit made for someone else, and he made several more before he decided it was time to start charging for more than just the cost of parts.[57] The first boat on which he made money was for Dexter Guillory, who convinced Benoit, who was tired of building boats by then, by insisting on paying for it.[58] When Benoit delivered Guillory's boat, he took Guillory out in a field to show him how to run it and how to cross levees: to show him "what the boat could do and what it couldn't do." Guillory's field was east of Eunice and was bounded on one side by Route 190. At one point during the orientation, Benoit looked up and saw thirty or forty cars pulled to the side of the road, stopped so their drivers could stare at this strange contraption, a boat that crawled through fields. (Greg Frugé might have been one of those who stopped that day; his shop was only a few miles away.)

Benoit himself eventually left the business, mostly because his heart was never in it, as a business. He remained a farmer and a fisherman at heart and never could imagine himself paying more than $2,500 for a piece of equipment that was part of such an uncertain enterprise as crawfishing.

~~~~~~~

Working simultaneously, two men took, what seemed to them, the logical next step in the development of the crawfish boat. A drive unit built around a hydraulic system made sense because hydraulics had become the means for distributing power or control over an increasing number of mechanical systems in agriculture and industry. Gone were the belts, gone were gears, gone were long spinning shafts with splines at each end. And good riddance, too, for clothing and body parts can get caught in any one of them. In contrast, hydraulic systems pose far fewer dangers because they distribute power within the narrow confines of the hoses, which are free to curve this way and that, with no direct line of sight to where the power must eventually emerge. To be effective, the fluid that runs within the hoses, and into and out of rams and through motors, must remain, well, within. The systems must be sealed. In being closed to the outside world, hydraulic systems are also more immune to, for example, dust and chaff that cause wear. (No system is truly immune; all systems are, in some fashion, open to wear.)

Both men, Ted Habetz and Harold Benoit, built boats for a time, but eventually they left the business for other pursuits, Habetz to focus on his chosen profession and Benoit to return to the farming and fishing that he loved. In doing so, the two ceded the field of crawfish boats to a group of makers who were, in some fashion, fundamentally different from them. These rising makers were men with dedicated shops, men who had trained in fabrication and welding and who were invested in making a place for themselves within the economy they knew using those skills. People of this sort have long been with us; we usually call them craftsmen, and it may be time for us to think of them as such again.

The first craftsman to make crawfish boats the focus of his business was Greg Frugé. Still affectionately known as "Momma Greg" to the larger community of makers, Frugé just happened to establish his shop in 1981, right as the boats were coming of age. Starting with the tiller-foot boats and developing his own chain drive, Frugé continued to build boats, under the fondly remembered name *crawfish combine*, until 2001, when other interests and other possibilities drew him away. During that twenty-year period, he built boats on a scale that only Kurt Venable has eclipsed, turning out seventy boats or more for six to seven years in the late eighties and early nineties.

Greg Frugé grew up outside Eunice, and it was perhaps only natural for him to set up his shop right there, on the north side of town, after he graduated from college. But, like Habetz, the schoolroom was not his sole source of education; before heading off to get his degree in industrial technology from what is now the University of Louisiana at Lafayette, he spent nine years welding and operating equipment. When asked, Frugé will happily tell you that he was lucky to grow up in a family where everyone thought with both their heads and their hands. His Cajun father was a rig welder in the oil industry who had his own equipment, and his uncles, on his German mother's side, were carpenters and builders. Growing up, he took his bike apart and repacked the bearings; he and all his brothers learned to weld at a young age. And, perhaps as important as anything else, Greg Frugé also grew up crawfishing.

Greg Frugé.

It was not long after he opened his shop that he found himself making tiller-foot boats and observing the enthusiasm and experimentation occurring in and around crawfish farming.

"Where before people who had ponds would just let you go in and fish them with buckets, they started getting smarter and started charging you ten cents a pound," Frugé said. "And we thought that was outrageous. You would fish your own crawfish and then pay ten cents a pound. It just sort of came about. And people saw where they could make money from this. It was all evolving. And I thought, *well, let me get in on the ground floor*."

Working closely with the tiller foot meant Frugé was also very familiar with the mechanism's tendency to break in its new application. The problem, as he saw it, was to achieve the gear reduction in a simple, and mechanically reliable, fashion.

"A lot of people were talking about needing to replace the tiller-foot drives, but no one made the step." His first thought was to stick with a mechanical setup to achieve the gear reduction. "Well, they were talking about different drives. And chain drive had come up in the past. Again, when you make equipment, most of the time it's not just your idea. It may be your idea, but it's from something you've seen in the past. And a drilling rig had what they called a compound on it, which is what they lifted the pipe out of the hole with. And it had this same type of drive. It had a chain drive, which was a gear reduction, to drive this big winch. It was fairly dependable. So I thought I would try it. Now, people did have chain drives, but they would run them out in the mud. And then you didn't have any way to reduce it twice. So you couldn't get the reduction that you needed. The wheels [of the boat] can only go so fast. You have to go slow in order to pick up the traps as you go."

None of the engines available then, and even now in many ways, can offer much in the way of torque, which is what you need to power any kind of wheel through the sticky mud of south Louisiana. You need a great deal of torque to power a steel, cleated wheel at something like walking pace through the sticky mud at the bottom of a rice field. The task is only made a little bit easier by having to power a floating craft. Torque is a complex thing, but maybe it is best thought of as the amount of work that the rotating shaft can do. A lawn-mower engine and a car engine can turn their respective shafts at a similar number of revolutions per minute, but how much work that spinning shaft can achieve is considerably different. (Anyone who has tried to cut grass that has grown too long and bogged their mower to a halt has had direct experience with torque.) Bigger engines provide more torque, but they also weigh more, take up more room, and consume more fuel.

One way to overcome the power-to-weight problem is to use small engines that can generate a very high number of revolutions per minute that can then, through gear reductions, be slowed down and, as a result, "torqued up." To do that, two calculations must be made simultaneously: how much work will need to get done (how much weight to be moved) and at what speed. Working backward, you eventually determine what size engine—which is almost always a question of how small you can go—operating at what RPM. The tiller-drive assembly had been a lucky historical accident that had opened one path. Habetz and Benoit had opened another with their application of hydraulics to the problem.

Frugé's chain drive was, in some fashion, a logical pursuit of the first path, and one that many people were interested in, since it offered an obvious mechanical solution, which almost every farmer could understand, unlike hydraulics, which were not yet fully a part of the agricultural landscape in the late seventies, and so there was, in some quarters, some suspicion of how well hydraulics would work and how well they would hold up.

Others had tried building their own mechanical solutions—what Ambrose Olinger once inventoried as "all kinds of jackleg contraptions"—but none of them held up well to the environment in which they were used.

"The mud and water would eat 'em up," Frugé said. "And that's when I came up with the idea to make a box with a double reduction."

Well, really a triple reduction. Frugé's original solution was built around an idler wheel that would engage and disengage the belt running to the box with the chain drive in it, but when Honda came out with an engine that had a two-to-one gear reduction built in, Frugé was able to get rid of the belt. That left the problem of steering the boat, which operators did trust to hydraulic rams. Frugé's response was to add a power-steering pump off a car driven, again, by a belt that came off the engine, giving the boat hydraulic steering. Dexter Guillory recalled that Frugé was into race cars, and that it may have been Frugé who first put a power-steering setup on a boat. According to Guillory's memory, at least for some of the first hydraulically steered boats, however the pump was powered, the steering only worked when you were moving.

"If you were caught in a corner, you were in trouble," Guillory said. "You had to move at least a foot or two before the hydraulic pump would start."

The chain-drive boats were, like the tiller-foot boats before them, push boats, whereas the first hydraulic boats were pull boats. This in itself was something of an advantage: as their power increased, the possibility of crossing a levee under your own power became more and more imaginable. The push boat's advantages here deserve a bit of explanation.

Levees in rice fields are narrow berms only a foot or so wide at the top, with shoulders sloping two or three feet to the floor of the cut below. One floor will, of course, be higher than the next so that water can be fed at the top of a field and then flow into each of the cuts, filling them to a predetermined depth before being passed on, via either a dip in the levee or a pipe or drain set in the levee. The shape of levees is, then, much like a double-sided ramp, and so, in theory, one should be able to drive up and over one. But physics gets in the way for a pull boat; that is, a boat with the drive unit at the bow. (One could call it *front-wheel drive,* but everyone with whom I talked only ever used the terms *pull boat* and *push boat.*) As a boat goes up one side of a levee, the drive unit itself goes up until it simply loses traction—if it were able to keep going it would, at some point, dangle from the bow. The net result is that pull boats cannot cross levees. The same cannot be said of push boats: they face other difficulties, but losing traction is not one of them.

The initial hydraulic-drive boats had front-wheel drive, and so Frugé's boats, with their chain drives mounted at the back, offered certain advantages over and above how power was communicated from the engine to wheel. But the days of the chain-drive crawfish boat were limited: building the drive was labor intensive, and the drive's mount on the boat took a great deal of finesse. The two together made the chain drive more expensive. However, in the process of mastering all these matters, Frugé had learned a lot about how to take full advantage of the boat's fulcrum.

This knowledge and know-how was significant, and it's revealed in an observation I heard from more than one builder who found some aspects of his boats being copied by people who wanted to build a boat for themselves: they didn't necessarily know what to copy. As Habetz experienced,

Frugé found people examining his boats, sometimes even in his shop, in order to copy the part numbers for the pumps and motors. They may have been able to buy the parts and assemble them into a boat that mostly worked, but they wouldn't necessarily understand the importance of the drive's mount point, and they would mount the swivel too low and the boat they built couldn't cross a levee. In some instances, Frugé recalled, they would come into his shop wanting to know why.

Another reason the hydraulic boat supplanted the chain drive was that the former could offer reverse, and so after a few years of making the chain drive, Frugé switched to all hydraulics, building both push and pull boats as customers wanted. He stayed interested in and committed to his boats crossing levees, and this resulted in a very curious design innovation: a kind of push-me-pull-you boat whose hull sloped upward at both ends, as if it had two bows, harkening back to the longitudinal symmetry of pirogues. While picking up traps, the boat operated like a pull boat, which many operators preferred since pull boats follow the wheel and need to be steered less actively, but when it came time to cross a levee to get to the next cut, the operator would turn the boat away from the levee, flip the drive into reverse, and push himself over.

Like Habetz and Benoit, Frugé entered the crawfish-boat business early, and he stayed with it for twenty years. By the late nineties, several things had changed. To some degree, he simply was bored with the business. As the form of the boat stabilized and matured and it really became a matter of refinements and not major developments, he found himself more interested in taking on new kinds of work, like vacuum-tank trailers. Adding to this was, from his point of view, the market was approaching saturation: almost everybody who wanted a boat had a boat. There were still new entrants, and existing operators would need replacements, but it was no longer the boom years when rice farmers and others suddenly realized that there was money to be made in including crawfish into their crop rotation. Finally, there was now a robust set of builders either focused on the crawfish boat or who had the crawfish boat as an integral part of the products they offered to their customers. (Business analysts might call this their "portfolio.") Thus, with the form well established, the market full or filling up, and plenty of supply, Frugé decided he was ready to do something else. As he noted at one point in our conversation, "A lot of people made their own boats, but a lot of people wanted a boat and didn't want to make one or didn't have the knowledge to be able to do it. And that's when I decided that this was a little something I could make some money with. That's one of the things you do in any business: find yourself a little niche, something that nobody else is doing."

CONSOLIDATION

A number of makers cropped up to satisfy the emerging, and expansive, demand for crawfish boats. Less like Habetz and Benoit and more like Frugé, they were mostly men who already owned shops. Almost all of them were fabricators by inclination, if not practice, and their diverse talents and expertise, as well as their own inclinations and obsessions, helped to create a dynamic, creative system that quickly took the front-wheel drive units bolted to commercially available hulls, moved the drive to the back, and came up with a hull capable of withstanding the impact of pounding up and down levees and wheels that made it possible to drive from one field to another. Anyone who has studied any kind of dedicated craft economy, as is found in this case, has encountered much the same kind of diversity. The diversity is, given human nature, probably inevitable, but it is also necessary for the ongoing existence of any tradition. Without the ability to adapt to the inevitable change that the world delivers to our door—perhaps more frequently than we like, perhaps less—no tradition survives long. In some instances, there may very well be more conservative as well as more dynamic practitioners within a given economy, but all that matters is that some of both dimensions are available to the system, and to the group, as a whole. Individuals may discount innovation, or they may embrace it. What matters is that both possibilities, and everything in between, are available.

The makers profiled here reflect larger trends: the first, Gerard Olinger, runs an agricultural-equipment repair shop; the second, Kurt Venable, operates a fabrication shop; and the third, Mike Richard, is a one-man welding business. All of these businesses bear the names of their owners: Olinger Repair Service, Venable Fabricators, Mike's Aluminum Welding. If you call, you will be greeted, first and foremost, with that name: "Olinger's" Gerard will say from wherever he is in the shop; "Venable's" Kurt's wife Sheryl will say from the office in front of the shop; and "Mike's" Mike will say, if he happens to be close enough to the telephone to answer it when it rings. Of the second generation of makers, Kurt Venable leads the way in terms of the number of boats made. Mike Richard makes about half as many. Gerard Olinger only makes boats now on order.

But there are others beside them: Dale Hughes is a young man whose manufacturing now includes not only crawfish boats but also agricultural equipment and industrial totes. Michael Quirk turns out metal hunting blinds that float as well as crawfish boats. Both men have a composite name to their businesses: Hughes Welding and Manufacturing, Quirk's Welding and Fabrication. Jimmy and Robert Abshire run Paul Abshire Welding Works in Kaplan. Like the Olingers, theirs is a brotherly combination, and their shop not only fabricates a small set of items, like water pumps, but also is regularly in the repair business—it's actually terribly hard not to be in the repair business when you are located in the middle of an agricultural community. Things break and, from a farmer's point of view, just need to get welded up right quick so that work can get done before the sun sets or the rain starts or the storm comes. If you have the means to bend or fix things back into position, then you will likely hear a large pickup truck pull up outside your door.

While such craftsmen dominated crawfish-boat manufacturing, there were always others. Some were regulars, farmers like Mike Cormier who regularly makes boats to sell to friends, neighbors, and acquaintances. Some are more occasional, men skilled in welding and fabrication, like Clayton Courville, who took a passing interest in making boats and then moved onto other things.

All of them contribute to the system's overall dynamism and efficiency, developing ideas or adapting ideas seen elsewhere. Some of them compete in various ways, and because boats currently can sell for $10,000 to close to $20,000, the money is significant enough to create enmity: I suspect such divisiveness is more a part of folk economies than folklorists would like to admit. Having brought it up, I should probably have a solution for it, or at least make it more clearly a part of what I document here. Suffice it to say, almost everyone I interviewed bore some grudge at some point in the past or present: something someone else had said or did. To raise those issues here would be to fall victim to the reality-television moment in which we now reside: a belief that interpersonal conflict and drama are more important than the things that keep us together. The fact is, there was, and is, a fair amount of money involved, and whenever someone undercuts someone else on price or seems to copy too quickly, too readily, someone else's innovation, there is going to be friction. Folklorists tend to turn such matters aside. For one, enmity usually passes and tempers calm. No one wants to have a passing remark move from conversation into the fixity of the written page.

Also not present in any of these accounts, especially of the active makers, are what they feel are their trade secrets: methods and designs that they have shown me or told me about that they feel give them a particular advantage, ones that make their boats handle in a particular fashion, wear in a particular way, or improve the speed or efficiency of making boats. It does not matter that other makers may already know, or suspect, something. If someone has asked me not to report something, it is not here. In the same way, after a fit of pique, I have been told, "Maybe I shouldn't have said that. Can you leave that out?" The goal here is not an exposé of personalities, but an exploration of how people collaborating, sometimes through competition, could so quickly develop a crawfish-gathering machine, a crawfish combine as Greg Frugé called it, of such amazing ability and complexity and also of incredibly reliability. How ideas travel across nodes in this dynamic network is discussed later.

~~~~~~

If you call the Olinger Repair Service, the voice you will hear announce "Olinger's" is Gerard Olinger himself, who is likely somewhere deep in the bowels of the shop where he and his brother Dale have worked together for decades now. In the days before he carried a cell phone and could see who was calling, you called the shop and left a message, and Olinger would return your call during one of several office breaks he would take each day in order to attend to precisely such things as telephone messages. He still retreats to his office to do paperwork or to place orders for parts he needs, but he spends far more time in the shop with this new arrangement.

Out in the shop, his time is split between working on current projects and receiving new work, which usually means spending time with whomever is dropping off the work to understand what they think needs to get done (which usually requires a translation into what actually needs to get done). Perhaps more than any of the other makers profiled here, the nature of Olinger's work is the most porous, a function of it being a repair business. Although he and his brother have a number of things they fabricate regularly—PTO ditchers and side plows, to name two—much of their work is in response to particular pieces of equipment in need of service or repair. The ditchers, while built on a common design with a small collection of templates associated with the design, are usually custom built, because farmers often have their own preferred widths for their tractor tires and ditchers follow behind the right tire. It's not uncommon to see a line of ditchers in the Olinger shop, each with a name written on it to make sure it gets to the correct owner. The same goes for the side-plow business; although Dale has put a plow on about every tractor ever sold in the area, that does not mean they all arrive in order. Typically, there are a good number of new tractors, but then there are a lot of older tractors coming in for a new side plow, and tractor manufacturers regularly change specifications such as mount points and axle housings from year to year depending on their own engineering and design requirements.

All of this means a regular flow of people through the shop on any given day. Each farmer or equipment operator or deliveryman requires a little bit of conversation, taking time that is sometimes in more ample supply and sometimes not. The Olingers have gotten used to dealing with such social nuances, and regulars to the shop also know when Dale is feeling particularly busy with little time to talk or when Gerard really does not want to immerse his hands back into a greasy ditcher assembly—the brothers know it, too, and gently poke each other about it now and then.

The intimacy of it all has a history, a reason for being this way: Gerard Olinger grew up one hundred yards from where he works today, in a house his parents moved to so that his father could help his grandfather farm the family land. The road on which that house sits, and on which Gerard and his wife Debbie's house sits twenty yards closer to the shop, is Olinger Lane. Its other inhabitants include Gerard's aunts and uncles, one of his sisters, and his brother Dale. Gerard is the oldest son of Ambrose and Margaret Olinger, who had six children in all, three boys and three girls, all raised in a two-bedroom, one-bathroom house that was one of only three houses on a road that otherwise snaked back into rice fields. Born in 1955 at the American Legion Hospital in the nearby town of Crowley, Gerard was eleven years old when he first began to help his father in "the shop."

Gerard's introduction to the shop was sharp and urgent: his father had broken his back when a loaded A-frame lift in the shop got its wheels caught on something and came crashing down on him. Because his father's mobility was severely limited, Gerard spent that summer in the shop, and he proved so useful that he spent subsequent summers and many afternoons after school there, too. It was his first job and, with the exception of a few summers spent working for farmers, his only job. There was never anything formal about the arrangement; it was just understood that after school he would get off the bus, grab a sandwich and a drink that his mother had waiting for him at the house, and continue down the lane to the shop. When he graduated from high school in 1973, he began working full time.

Gerard Olinger moved out of the Cove in 1975 after he and Debbie were married, and they moved into a house in Rayne that belonged to her parents. All three of their sons—Mark, Matt, and Paul—were born there, but they, like their father, did most of their growing up in the Cove because in 1985, Gerard and Debbie built a house between his parent's house and the new shop. The old shop no longer stands, but it had once been his grandfather's equipment shed. Gerard's father, Ambrose, had slowly converted it to a repair shop over the years.

Ambrose Olinger had intended to be a farmer like his own father, but as his reputation for repairing machinery grew, and more and more farmers began to line up outside the shed with equipment to be repaired, the shed slowly became Olinger's Repair shop. By the late 1970s, with Gerard now working full time in the shop and Gerard's own expertise as an extraordinary machinist becoming established, the father-and-son business had outgrown the converted shed and they moved into a new, custom building one hundred yards southwest of the old one.

In 1987, Gerard Olinger took over the shop, buying out the business's inventory and paying his father rent on the building. In 1989 he was joined by his younger brother Dale, who had been a successful manager at a bank branch in nearby Church Point until its parent bank failed and he lost his job. Dale's availability came at a crucial time for Gerard, who was busier than ever, but his longtime partner, his father Ambrose, had recently retired. Dale agreed to come help out but insisted that he didn't intend to stay for long. Gerard responded that as long as he wanted to work, he had a job, and they "haven't talked about it since."

Olinger's entry into the crawfish-boat business came early, and was, to some degree, a natural outgrowth of the shop's focus on agricultural-equipment repair and fabrication. While there is always a steady stream of repairs, the Olingers—first as Gerard and his father Ambrose and then later as Gerard and his brother Dale—have also long turned out a variety of tools, machines, and pieces of equipment that are integral to local farm operations but are not otherwise available, or are not available in forms or materials suited for the harsh environment of the Louisiana prairies.

The first boats were built in collaboration with Jerry Heinen, who is not an atypical farmer of the area: Heinen can calculate the amount of rain needed to raise any given field you point out by whatever height you call out. He is that good at math. And he is not alone. Farmers regularly calculate acreage of cuts and fields over and against the amount of water they need to pump to flood or because they need to anticipate fuel demand, and thus fuel costs, for the month. Math is intrinsic to the work. Jerry Heinen is simply someone who enjoys math for math's sake. (A more comprehensive examination of farming in the area would reveal as many differences among farmers' abilities and temperaments as there are among the boat makers, and that those differences are crucial to the dynamism of the overall ecosystem of ideas and practices. And farming in south Louisiana is only a subset, and thus probably a decent index, of American family agriculture in general.)

With Heinen's prompts, Olinger turned out his first boats, which looked a lot like everyone else's boats: the hulls were closely modeled on commercially available hulls, which resulted in a fairly crowded boat. It also resulted in a boat that had some difficulties in keeping water out of the stern when it crossed a levee, and so one of the first adaptations that I have seen, but Olinger himself does not recall, is the addition of small fences at the stern of the boat to give the hull a bit more height.

As the eighties wore on, Olinger began to build wider boats, which gave them a bit more flotation because their displacement had increased with the width. The wider boats not only handled better but they also opened up the interior of the boat for the operator to move about more easily, and, with the increased flotation, to stack more crawfish in the bow. But the singular innovation, and one to which I have to attribute to Olinger, is the conversion of the crawfish boat into a truly amphibious vehicle. No one else with whom I spoke could recall who first put wheels on the front of a crawfish boat, and Olinger himself is reluctant to claim credit for the idea, and yet, when I asked about it, he recalled doing it and why.

"Farmers kept bringing their boats back with the bottoms wore out. It happened again and again," he said. "So I finally asked one of them what

Gerard Olinger backing a crawfish boat into his shop for repair.

was going on. They were going down the road with the thing. Just plowing along a road with it."

Most of the roads between rice fields are dirt lanes, but some have gravel on them, which could hardly be good for a sheet of aluminum bearing hundreds of pounds of man, machinery, and crawfish. There is also, of course, the blacktopped highways that run throughout the region, which are also used by farmers to get from one field to another. The result was an incredible amount of wear in a short amount of time, something that at first perplexed Olinger until he hit upon the solution.

"So I figured as long as they were going down the road, I'd give them wheels."

The rest, as they say, is history.

With the addition of the two front wheels, typically small trailer wheels but in some cases larger ones, the crawfish boat had reached full maturity as an indigenous amphibious vehicle capable of moving across the Louisiana landscape, dipping into one rice field to retrieve hundreds of acres of crawfish by clambering from one cut to another, and then driving up and out of the field and down the road to the next field.

The move was both audacious and obvious. No one looking back imagines it otherwise: of course the boat got wheels. The innovation was quickly adopted, with each maker adapting the innovation according to his own theory about the way boats were supposed to look and operate. Olinger's first impulse, and the practice he maintained for twenty years, was to set the wheels into wells so that the overall lines of the boat were kept clean. When Kurt Venable started building boats, he followed much the same logic, and to this day his wheels are placed internal to the hull, dropping out of the bottom of the hull from two boxes inset from the sides, almost as if they were landing gear on an airplane tucked up for flight. The rest of the builders did not follow this practice: Mike Richard bolts his wheels, which are fairly small, to the sides of his hulls, with a steel pipe stretched across the interior of the hull for reinforcement. Indeed, Olinger himself, tired of the stress fractures he was constantly repairing, eventually took to bolting the wheels to the outside of the hull, noting that the wheels do not, as he once feared, get in the way of most operators, nor do they affect the boat's overall handling characteristics very much.

Olinger continues to build crawfish boats now, though typically only a handful a year, and in some years none at all. He does a fair amount of repair work on boats, where they join grain carts in a line of equipment to be repaired and/or refitted. I have seen, for example, more than one hull from a used Venable boat come into the Olinger shop so that an Olinger drive can be fitted to it—the different styles of the drive units is perhaps where variation is at its greatest among the makers and something to be considered next.

~~~~~~~

To hear Kurt Venable tell it, he is merely an observer of the development of the crawfish boat, a simple maker of sturdy craft who has added to the overall economy of ideas and objects. In reality, nothing could be further from the truth, since, as everybody knows, Venable has long been the most dominant maker in the market, turning out around 1,400 boats in the past three decades, perhaps more than all the other makers combined. Standing in the only air-conditioned shop that I encountered, Venable is a man both proud of the efficiencies that he has introduced in his own process and still somehow shy about his role in the overall history of the crawfish boat.

Venable Fabricators is on the western edge of Rayne, right on Highway 90, in a building that used to be a tractor dealership. The main building is about sixty by a hundred feet, just a bit bigger than the Olinger shop, and contains a variety of CAM (computer-aided manufacturing) equipment as well as more traditional pieces of hand-operated equipment. Off the back of the building is a shed roof that gives Venable and his employees, which have ranged from one to three during the years I visited, a place to work that is out of the weather as well as shaded from the sun. Also off the back of the shop and open to the shaded work area is an enclosed room that contains a CAM-controlled plasma cutting table. There are a few more smaller outbuildings that take up the rest of the lot on which the business sits, as well as a parking area for boats that have already been or are in line to be repaired.

In a front corner of the main building is the office, where his wife Sheryl, herself a Leonards from Roberts Cove, keeps track of just about everything and everyone that comes into and goes out of the shop. There is an easy give-and-take between the two: when he cannot remember something, she can place the date—sometimes they triangulate by referencing various events in their combined lives. Whereas he started the business and ran it on his own for the first two decades, she has clearly become central to the business as it is today. If CAD (computer-aided design) drawings are usually open on his computer, it is bookkeeping software that dominates hers. There is an endearing pride in his voice as he describes her importance to the business.

To be sure, business, and life, have not always been this good. One of the things that Venable makes clear in the history he tells about himself and the business is just how much he struggled in his first decade of being on his own to secure a steady, dependable income for himself and his family. First working for his father, then his brother, and then living through some wobbly moments in the early years of being on his own, it is only when he solidified his business around the crawfish boat that things fell into place. Given this, much of his conversation is about the business dimensions: product design, workflow efficiencies, and inventory control. But you can't help but notice that constantly peering at you through such talk is his real love of how things work. Anything that gets him closer to that is worth the effort or the investment.

Some of things that do that are the CAM machines that are placed about the shop. The way Venable tells it, he had to buy the machines because of the money they have saved him, and that is no doubt true. It is also conceivably true that the machines are only a natural extension of himself as a layout guy, as he calls himself, drawing on a term from the sheet-metal industry: someone whose imagination is fundamentally three dimensional and who values precision above all else. In Venable's own telling of his story, it was his foundation in fabricating complex, three-dimensional objects as a boy, and later a young man, in his father's shop that led him to where he is today.

His introduction to metalwork was in many ways inevitable; his grandfather Guidry and his father Logan Venable both owned and operated sheet-metal shops. His grandfather's shop was in Crowley; his father's in Rayne. Venable remembers working afternoons and summers in the shop, playing quite a bit but later getting some real work handed to him. In junior high and high school, he and a friend would scrounge his father's shop and the shop of the nearby rice mill for spare parts to build things like a minibike or a go-cart. "We were just terrible," he remembers. "Flat-out dangerous: because the minibike, for example, didn't have a clutch or a brake. You cranked it up and it just went. Oh, we liked to killed one of the guys. He got on and there was no way to stop. I thought his mother was going to kill us."

Much of his father's work was in fabricating and installing metal ducts for the newly arrived, and much-prized, central air-conditioning. Venable remembers that most of the shops in Rayne came to specialize in doing air-conditioning work and not much else. It was not, given his experiences of crawling in cramped, hot attic spaces, work that interested him much, as much as he liked the machinery involved. The fabrication work was done mostly with thin-gauge galvanized sheet metal that was either brazed or, later, when the work began to pile up, bolted together. Sometimes they would still get a chance to build a chicken-feed trough or something else for a farm, but it wasn't until the Doré Rice Mill burned down in 1974 that the work became interesting to Venable.

That summer and the next he worked a great deal in his father's shop rebuilding the blowpipes that moved the rice about the mill: some pipes carried rice up and onto a conveyor belt, other pipes took the grain to a bin, while still others took away the chaff. Venable's father's shop still contained all the templates for making the various pieces of pipe, and the work demanded the kind of intensity of focus that seems to drive Venable.

Eventually, Venable started his own shop in Crowley; wanting to focus on a product around which he could build a business, he developed his first product, the Cajun microwave. There are a couple of things to know about the Cajun microwave: it is a folk name for an artifact that, like the crawfish boat, is widely fabricated. The *microwave* appellation is, of course, a humorous allusion to the actual microwaves that were becoming more widely available and to the penchant that many natives of the region feel that the Cajuns have for making something for themselves. The joke is based on the notion that Cajuns, historically, were French speaking and thus likely to misunderstand something that came from outside their environment, typically English, and that Cajuns, again historically having fewer economic resources, were often either too poor (or cheap, depending on who's telling the story) to buy something and preferred to make it for themselves. The term is applied rather widely, and it is not unusual to find someone digging a hole in the ground and lining it with sheet metal in order to cook a pig and calling the resulting fire pit a Cajun microwave.

The Cajun microwave as built by Venable was a version of an old-fashioned Dutch oven, an iron pot with a tight-fitting lid that has a rim on top to hold coals. In this way, heat can either come from below and above or only from above. Venable took the idea of heat coming from the top and built a specialized box that, made out of stainless steel, was easy to maintain and use. His specific inspiration was a Russian army camp stove, but he also developed a number of improvements, which he patented. The microwaves were a hit, and he was considering expanding his fabrication abilities when the venture hit a serious snag: in seeking to expand the retail presence of the ovens, he managed to convince a big-box store to carry the ovens. Unfortunately, the giant retailer sought to capitalize on the popularity of the ovens by making them loss leaders and selling them at cost. It was a lesson that many small- to medium-sized businesses have had to learn when they find that their popular product is useful to others in unexpected ways.

"When I went into business, it was a terrible fight until 1984, a fight to stay alive. The oil field collapsed. We were making closures for platforms that were insulated to keep equipment from freezing. All those things that were available to do, that all evaporated."

Reese Guilliot came to the shop with a particular request: he wanted a ten-foot-long by four-foot-wide aluminum boat with front and back decks on which he could mount a ten-horsepower engine to crawfish. Venable doesn't remember many people welding aluminum in the area: if you wanted a boat, you had to go to Moreau's Marine and buy a DuraCraft. The problem with the DuraCraft hulls was that they were only three feet wide, too narrow to float in nine inches of water when fully loaded. For Venable, fortune smiled on him that day: he had just found the kind of work that really challenged him. Because he had learned how to design parts for building flow pipes and had constructed elaborate drawings in service of submitting his patent on the improvements he had made to the Cajun microwave, building boats was, to him, a logical next step: "I was a layout guy, so it was easy for me to go into boatbuilding." He bought books on lofting and read them on vacation. He then came back and built a lofting floor in his shop and began to take orders for boats.

"I wanted to become an aluminum boatbuilder because I would have no competition," he said. "Nobody does that. Well, you have your big guys, like Breaux Brothers. I decided I would come to Rayne, set up in my daddy's shop, and build boats."

In 1984 Venable sold all his equipment and moved to Rayne. "I'm done.

A Venable boat sitting next to a rice field. Note the trap, lying on its side, and the sacks sitting atop the bow deck.

I don't need a press break to make boats. All I need is a Skil saw, a Heliarc machine, an actual MIG wire feeder, and a chop saw, and I'm done. I can build you any kind of boat you want."

Venable's confidence came, in no small part, not only from his experience with sheet metal but also with his love for boats. From as early as he can remember, his mother would take him and his siblings—and with eight children in all it was quite a family—to the family's camp at Toledo Bend while his father stayed home to work. Weekends and summers, the Venable children would take a twenty-five-horsepower outboard in the back of the family car and rent a boat once they were at the lake. The summer after eighth grade, Venable remembered, he and his two sisters were fishing in a cove in Toledo Bend when the outboard engine wouldn't start. He had to swim until he was close enough to yell for help, so they could be towed in. Experiences like this only seemed to have encouraged him.

"I could do whatever you wanted," he said about the possibility of making boats. "I was sheet-metal oriented. A regular bateau is a sheet-metal object. Because of my experience, I wasn't scared of things like rivets."

"If you are a true sheet-metal guy, and you are good around boats," he continued. "Well, let's leave boats alone. Let's say I know sheet metal. Somebody will come to me and they have a simple icebox behind their truck. They tell me they need one like that commercially. Can you help me? Of course. You go find out what kind of unit you need for the size box they want, how much meat they are going to put in there. Then you build a box. You cut the holes through the wall for the unit to fit. You build things. It doesn't matter if it's a boat, a cooler, or a funnel to cut a duck's head off. You look at things. I look for structural members. I look for the thickness of the hull."

He had learned a lot during his years in Crowley. With his shop across the street from Moreau's Marine, and much of the workflow of the Cajun microwave taken care of, Venable was often called upon to do a variety of repairs for Moreau's. During those years, he would listen to what people would tell him about the way boats were built and why. If a boat came in with extra-thick hull plating, he would learn why. "All those years you listen and it's all in your head."

For Venable, it all comes back to sheet metalwork. If someone came in and wanted a cabin in the boat, he would use, as he calls them, his layout skills to measure openings, calculate the radius of the corners, and determine the width of the gap to be filled by the gasket. And he was always open to opportunities. When another client of his steel dealer went out of business, the dealer asked Venable if he was interested in some of the materials that had been intended for the other client. Sure, he replied, make me a deal. They did, and he found himself with a ready supply of steel sheets to make a variety of bass boats.

Venable did distinguish between "true boatbuilding" and "sheet metalwork." Bateaus and bass boats seem to be simply, to his mind, sheet metalwork. Lofted boats are different. How did he learn to loft? He taught himself. He bought a book.

"You have to understand: I'm a layout guy," he said. "And what you don't understand about a layout guy is that we draw things in plan view and profile view. And we look at it. And we pick our highs from those views to make, for example, reducing offset funnels. I can take a flat sheet of metal and make you any type of fitting you want. I could start off with a twenty inch on a forty-five-degree angle and end up with a two inch on a ten-degree angle. Lay it out, snip it out, roll it, put it together, and here's your fitting. That's layout. Boatbuilding is easy compared to that."

The move from layout to lofting seemed like a logical step. When he saw the photographs in the book for the first time, he realized they were doing things he had long been doing out of necessity and his own drive to work precisely. His course, to his mind, had a certain inevitability to it. It's a larger inevitability. "Anybody who builds boats is going to eventually evolve to that state," he observed in reference to the Vikings' progression to lofting.

To build a boat, he would build a small model out of sheet metal, a half-size model. He started with a keel, and then he added a chine, then a bottom, then a gunnel. Then he had a boat. He shaped it until it looked like it was supposed to look, then sliced it into sections and laid those sections out. But there are bateaus and then there are vessels. The difference, for Venable, is critical. The former are flat-bottomed craft that are going to float, and not much more. A vessel can withstand the forces of the open sea. "If you took a flat-bottomed boat and ran it into waves, it's going to crack. If you take a boat that's shaped like a bowl, you can't crack it. Because now it's a vessel: it has no flat surfaces." He follows that up with an obvious example: "You can't pressurize a square tank. You can take a cylinder, put some ends on it, and blow it up to two thousand pounds and it won't break. You take a square box and you can't even achieve two PSI."

And the transition to crawfish boats?

People simply started asking him to build a boat for them. He started with small boats, five feet wide by twelve feet long. He had already been doing some repair work on other people's boats, so he knew what did not work, but he also knew that he could not make a boat too heavy. He knew how to calculate buoyancy from building fishing boats, so when someone told him they wanted to operate a boat in nine to eighteen inches of water, he was ready to make the calculations: at five feet wide, with a sloped bow, he was able to determine that the boat would draw seven inches. He wasn't sure what would go into a boat, but when he suggested building a fourteen-foot-long boat, his first customers weren't interested in the additional expense. Actual use, with mud building up and everything else that gets dumped into a working boat, proved his estimate off by two inches. The boat drew nine inches to float.

Venable's first boat was for Harvey Thibodeaux in 1984. He built the hull and Ambrose Olinger built the drive unit. Another customer got a hull made by someone else, Kaliste Leger, and went to Venable for the drive unit, a tiller-foot push unit. Venable's first complete boat was for Thibodeaux's brother, who came to Venable and wanted the entire setup, hull and drive unit, made for him.

By 1986, people were coming to Venable wanting hydraulic boats, and they wanted to be able to go over levees in the boats. His first step was to go around, examining what everyone else was doing, trying to understand what worked and what didn't. But he didn't always have to leave his shop to do that: boats were also coming in for repairs, and that allowed him the kind of thoughtful examination that seems to make him most happy. One boat came in that had a hydrostatic drive system on it. Venable again studied the situation, and then, as he noted, "we lucked out."

At the same time, he knew that the Olingers were building a pure hydraulic boat with bigger pumps and motors. While Ambrose Olinger was unwilling to let Venable learn anything, Gerard and Kurt had long been willing to share information. In fact, Gerard Olinger was willing to sell Kurt the parts he needed. Venable built four hydraulic boats in short order.

At the same time that Venable was beginning to build boats, he had

A Venable boat at work in a field.

already started building vibrating graters for crawfish boats with Buck Leonards. At the 1987 trade show hosted by the Crawfish Farmers Association, they had a sample grater and a video of it in action. Both were excited because one of the choke points in crawfishing was how fast crawfish could be sorted as a boat made its way through a field. While some graters struggled to keep up with two pounds per trap, Venable and Leonards had their grater easily handling eight pounds per trap—they had, after all, put that much in each of one hundred traps in a field for the purposes of demonstration. The video was impressive: the traps were so heavy with crawfish that the teenaged Billy Link, who was the operator in the demonstration, simply hove the mounds of crawfish onto the tray, and the grater just shook all the crawfish into the waiting sacks. And then, in the video, the boat came up to a levee, and this was important to Venable and Leonards because they wanted to show that the grater was not going to turn over when the boat pitched up and then down again as it went over the levee.

When the boat went up and over the levee, what people wanted to know was "who in the hell made that boat? That's what come out of their mouth." Venable replied that he had made the boat, but that wasn't what he and Leonards did and it wasn't what they were trying to sell. Although the two of them may have wanted to start a grater business, Venable left the show with twenty-five orders for a boat.

They sold twenty-seven boats the year after the show. Some were equipped with hydrostatic drives, some with gear pumps. By 1991, Venable Fabricators was making thirty boats a year. The basic shape of the hull was established early, and Venable has worked hard to make only small changes. One example of such a change is the lip now at the rear edge of the bow deck. It was added when Venable saw how much water the noses of the boats were picking up as they plunged into one field from another. He added the lip in an effort to shed the water to the sides of the boat. His solution, he observed, was that of a "sheet-metal guy": he added a series of bends that takes one sheet of aluminum up at a steep angle and then vertically on the back to wrap under again for a smooth finish that also adds strength to the hull.

The hull is the key to his boats, and Venable credits a wooden mud boat built by L. C. Touchet as his inspiration. Someone wanted him to build an aluminum version of the boat, and as Venable examined it, he realized that every single surface was curved, even the transom. The boat's owner, Kenny Ancelet, told him that it ran well with only a fifty-horsepower motor, and that what he wanted was an aluminum version that wouldn't require as much maintenance: "It's getting too old, and I'm tired of painting it." As Venable looked at the boat more closely, he began to realize how much sense the curves made. For example, the middle of the boat, its belly, was deeper than the transom at the back of the boat, which gave the boat a way to release the water as it flowed underneath. And, of course, every curve added strength: "You couldn't bend that transom with all your might, because of its shape.

"It was a true boat. There's a definite advantage to a vessel shape."

Venable took notes on everything he saw and applied it everywhere he could in his own boats. Three decades after building his first boat, he is still making boats, although now he oversees a shop run by his son and staffed by young men his son's age. The crawfish boats are central to the family's business, and they are central to the boat industry.

BUILDING A BOAT

Right now, the man who makes the second-greatest number of boats in most years is Mike Richard, whose shop is in the community of Richey, just south of Eunice. The shop and the house he shares with his wife Susan, who is also his bookkeeper, practically run together, with the occasional sheet of aluminum leaned up against the house as he moves material around. Out of those aluminum sheets come hydraulic boats and their smaller cousins, the boats pushed by hand, as well as aluminum weirs and other bits of gear that rely upon his four decades of experience with aluminum welding.

Crawfish-boat manufacture and repair occur year round, with new-boat construction kicking into high gear somewhere around the end of August or the beginning of September. Richard continues to build for as long as he has orders, usually making a dozen boats or more in any given year. He typically finishes in time for crawfish season to start in January or February, turning then to the boats brought in for repair that have already begun to line the yard in front of his shop. He will spend the spring emptying the lot.

Richard started building boats in 1985, when he left Kaiser Aluminum. Like others, he began with the tiller-based design, and his recollection is that he stayed with some version of that until he switched to hydraulics in 1992. He had built boats that used hydraulics before that year, but he was just building the hulls, not the drives, which customers then took up to Frugé's Cajun machine shop to be outfitted with hydraulics.

Richard's is a one-man operation, and so it is perhaps easiest to see a boat being made if we linger in the yard in front of his shop. With only one man's action to narrate, it is easier to follow the action. Because there is no simultaneous work, an observer can have some sense of seeing everything done in the sequence in which it gets done. Richard also keeps his preassembled materials to a minimum, preferring instead to build the necessary parts as needed out of stock material. That isn't to say that there aren't some premade pieces in his process; in addition to the engine and the hydraulic pumps, pistons, and valves that he must purchase, there are a handful of pieces cut from half-inch-thick steel that he prefers to buy already cut and that he keeps stacked neatly in various corners of his shop.[59]

In brief, someone setting out to build a crawfish boat faces one fundamental challenge: the transfer of power from the boat's drive unit to its hull. Unlike other boats, where perhaps the forward thrust of the boat is somewhat cushioned by the fact that the propeller, or paddle, is pushing against water, the crawfish boat's thrust is achieved by pushing against land: the vanes that stand out on a boat's wheel are not paddles that dip into the water but rather cleats that dig into the earth. Any and all starts, stops, changes in direction, or bumps are transferred immediately, and proportionately, to the hull. This vital transfer occurs at the intersection of the steel drive unit and the aluminum hull. In addition to the jolts and jerks of the drive unit that the hull must absorb, it must also have a systemic way to handle regularly being crashed onto and over levees and roads.

In every instance, each maker has his own theory about the nature of the forces at work and how to counter, or redirect, the force successfully—success here is gauged by how long a hull lasts before it develops cracks in the aluminum significant enough to require repair. Braces, gussets, reinforcing bars, and channels: all are used in some combination to fight off the two agents, one big and one small, that work on any material object that sees use. The big agent is trauma, and it's found in the kind of impacts just described. The small agent is wear, and it is found wherever two surfaces come in contact and either must move or come to move through some indirect action. Where movement must occur, there is generally anticipation of wear, and a variety of bearings and bushings and blocks are used to counteract the slow dissolution of seemingly solid objects through friction. But wear occurs other places as well; I have seen half-inch divots in aluminum railings from where a hydraulic hose rested. The hose may have only shifted a half-inch in either direction as it pressurized, or depressurized, to feed a hydraulic piston. But think: a half-inch a hundred times in a day times a hundred or more days in a season times a half-dozen seasons. As Dale Olinger once observed, as I helped him work on a combine header, "We are getting rid of the time. There's paint, and there's rust. Then there's time."

Accounting for time and for users, the boat makers have arrived not only at a variety of ideas that can be seen in the manifest form of the boats themselves, but also in a variety of construction processes, some of which they regard as shop secrets; yet another reason for not providing too detailed an account of the making of a boat here is that in every instance of being shown something and being asked not to tell about it, I have honored that request.

The account that follows is thus specific enough to be understood as the particular process of one maker but also general enough not to reveal any of his particular secrets as well as generalizable to the larger field of makers; each of them confronts similar, if not exactly the same, problems and produces his own solutions based on his own ideas and experiences. That one came from a welding background, another from a farming background, and another from a repair background only reinforces the diversity of ideas and practices to be found even in this rather small network of craftsmen.

In most years, sometime in August or September, depending upon the number of orders, Mike Richard will turn from other kinds of work that he takes on, such as aluminum weirs or other types of water-control devices, and begin to assemble boats out of the pile of aluminum sheets that sits outside the front door of his shop. During the course of a single week, he will single-handedly bring a boat—hull and drive unit—into being, drive it out of his shop, and park it under the shade of a tree to await its owner.

Like most of the boatbuilders, Richard begins with the two pieces of sheet aluminum and one piece of aluminum angle that when tacked together with a few welds make the hull magically spring into being. He uses a sheet of tread plate, which most of us recognize from metal stairs we have climbed, for the bottom and a sheet of regular smooth aluminum, cut down its length to form two pieces eighteen inches wide, for the sides. The sides are as long as the boat, typically about fourteen feet for Richard's boats, and the bottom sheet just a little longer, needing the additional length to give the boat a graceful, curving front end. The angle material is as wide as the bottom sheet and will eventually help to shape the boat's nose, or bow, into a taper.

The first bit of construction is to form the bow and from this—the front end of the basic box of the hull—everything else proceeds. Richard

Mike Richard measuring from the front of a hull to establish where the axle will eventually be and where the hull will assume its full width as it flares from the bow.

first tacks the piece of angle stock, facing down and inward, to one of the sides and then to the other. Then he draws the flat sheet of aluminum lying on the ground upward and tacks it to the angle stock. With the four initial pieces tacked together, albeit loosely, forming the bow, Richard first reinforces the welds on the inside and then tacks first one side and then the other side into place, leaving a bit of the bottom protruding beyond the sides. Later, he will use this small lip to add a weld on the outside of the seam where the sides and bottom meet in addition to the weld on the inside of the hull. For now, the tack welds are an inch or two long, spaced between eight and twelve inches apart, depending on how the sheets are behaving. There are, for example, more tacks where the curve of the bow straightens out into the flat of the bottom.

To create the slight flare out of the sides of the boat, Richard measures four feet back along each side piece and marks the placement of his stretcher bar, a piece of square steel tubing with notches six feet apart. It takes a bit of persuasion to get the straight pieces of aluminum sheet to form the graceful curves, but a hammer in the right spot makes it happen.

With the sides flared properly, Richard attaches the top piece of the transom, using some leftover tread plate, or drop, which he has cut to the angle formed by the sides. After a few tacks on the outside, he gives the inside a thorough, continuous welded bead. He has to, because his next step is to weld a piece of two-inch by six-inch aluminum channel on the inside of the transom plate. This kind of reinforcement is critical and is only the first of many such reinforcements. The transom is where all the force of the drive unit is transferred to the boat, and it is critical not only that the transom can withstand that force, which wants to crush it like an aluminum can, but also that it can successfully transfer that force to the boat so that it is able to move effectively through the water and over land. In the later stages of Richard's build, we will see the other pieces of this transformative apparatus that he, like the other boatbuilders, has developed over the years.

With the hull effectively closed in—the bottom, angled part of the transom is not yet in place—Richard now lays in a continuous weld on the inside where the sides meet the bottom of the boat. It's hard not to admire a man in his fifties who so easily folds himself up in order to get close to the work. On his hands and knees, with his thighs close to his chest, he is also completely covered, despite the August heat. He has on a

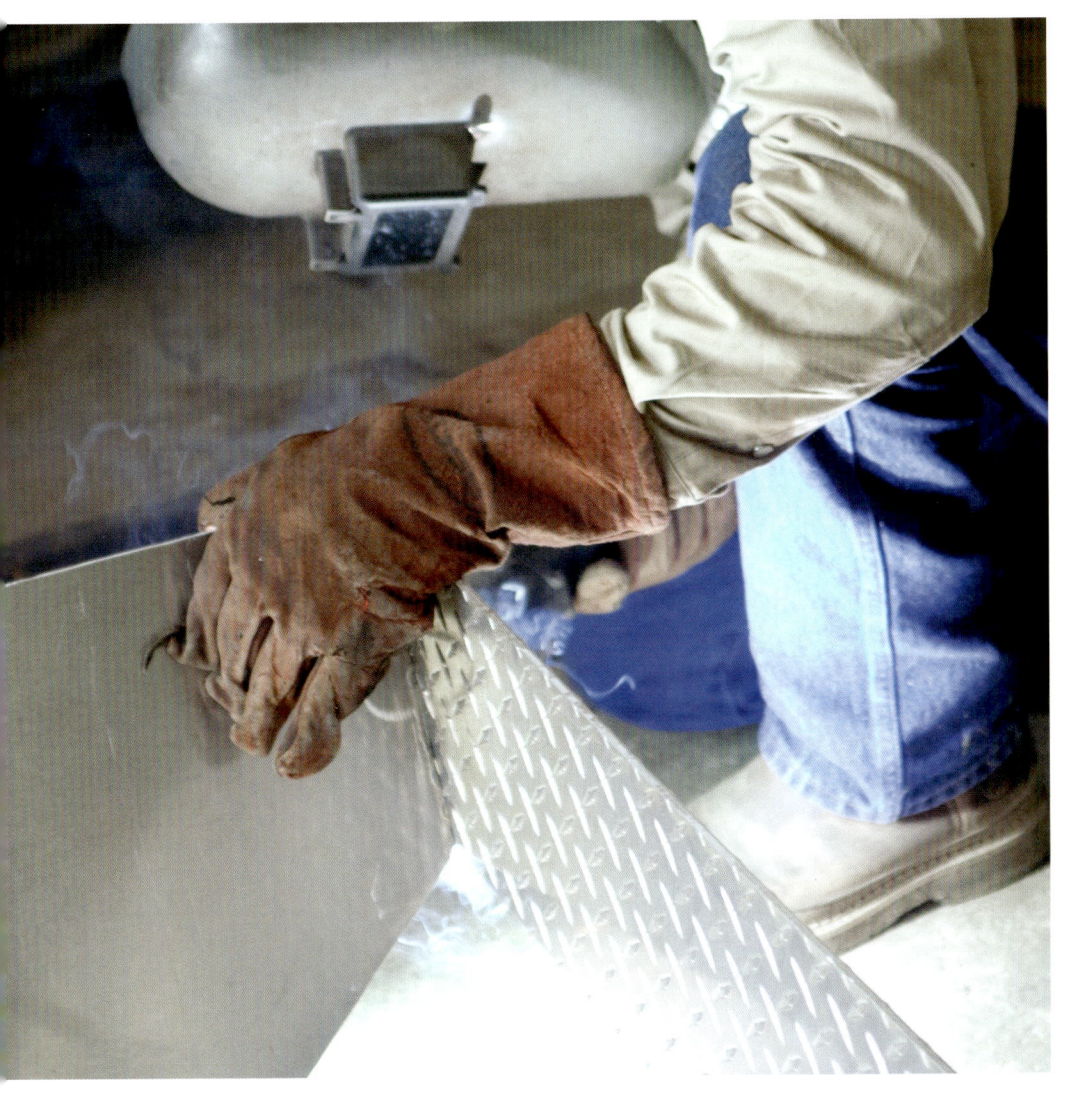

Richard welding the various pieces of the transom together.

heavy cotton shirt over his T-shirt and a good pair of jeans over his high, laced work boots. There are gloves on his hands and a welder's cap under his mask. He exposes nothing to the dangerous light of the welding arc, which is reflected across the myriad bright aluminum surfaces of the box in which he now crouches.

Richard again measures back from the bow, but this time only a couple of feet. He then measures across the width of the boat, at both the top and bottom of the sides. His measurements describe a trapezoid that he then transfers onto a drop from the sheet of smooth aluminum. The complete quadrilateral need not be drawn: he halves both measurements and uses them to mark a center on the drop. With the piece cut and then bent twice along its top edge to form a channel that not only makes for a nicer and safer lip for the equipment box that he is creating but also gives the boat greater structural strength, Richard notches the bottom of the piece to accommodate the welds already in place along the bottom of the hull.

After he has tacked the bulkhead into place, Richard grabs two pieces of gunnel rail and places each one in turn on a side of the boat and marks where the hull tapers to the bow. He uses an acetylene torch to heat the rail and bend it, by eye, to the appropriate angle, never failing to get it right. He once again dry fits the now-bent rail to the boat and marks where it must be notched to accommodate the lip of the bulkhead. He makes the cut, freehand, with a circular saw and then fits the gunnel fully onto the edge of the sides, using a pipe wrench for leverage and an occasional tap of the hammer to make sure everything fits securely.

With the gunnel welded into place—the welds here do not need to be continuous and so the work goes a bit more quickly—Richard does one of the most amazing things I have ever seen: he cuts end caps for the gunnels out of a piece of aluminum using a handheld circular saw. This might not seem virtuosic on its face, but consider that the fillips of aluminum are shaped like the letter P, a P that is only two inches high and wide and whose descending base is only about a quarter of an inch long. With my jaw open (and catching the occasional bit of stray aluminum), I can see a slight smile on Richard's face as I ogle at the sureness and deftness with which he handles the precise cuts. He uses an old plastic crate to prop up each side of the boat to make it easier to weld the fillips onto the ends of the gunnel rails. Afterward, he grinds everything until the capped rails are shiny and smooth to the touch.

Richard's amazing freehand cutting of aluminum with a circular saw.

Richard next single-handedly flips the hull over, picking it up, sliding the edge back through a careful walking back of each end, and then laying the whole thing back down again. With the bottom of the boat exposed, his first task is to finish closing in the hull by adding the bottom half of the transom that also angles in, a unique dimension of crawfish-boat hull design that keeps the boat from digging into levees during crossing. Richard utilizes more tread plate for this and uses a crowbar that he holds with his left arm and the left side of his head to keep the seam between the two pieces of transom tight.

With the hull still upside down, Richard now needs to add the three pieces of aluminum angle stock that run longitudinally along the bottom of the boat. Although these pieces may give the boat some lateral stability in the water, their chief purpose is to absorb the wear and tear that would otherwise be dealt the boat's bottom, angle stock being more easily and cheaply replaced. Richard cuts the three pieces to length and then trims the front of each piece into a deep V, making them look an awful lot like very unwieldy spears. He then tacks each piece into place, following marks he has already made that give them the spacing he prefers. Because these chines start up along the slope of the bow, the twelve-foot-long pieces of stock need to rest somewhere while he welds the tips in place. Richard smiles as his secret shop weapon emerges, an old metal folding chair, and he places it expertly at exactly the right distance from the bow to give each piece a place to rest as they angle up into the upper recesses of the shop's interior.

All of the builders have their secrets, either materials used or methods. Some have a supply of jigs or templates; some have specially designed tools or have modified commercial tools for a particular purpose or a particular stage in the construction of a boat. Some of the secrets are hard won from experience, and some of them are a function of a builder's particular design and/or fabrication philosophy. People who aren't makers may not necessarily understand that there can be philosophical differences in welding, but there are. In the same way that different tactics, or tactics in different sequences, can reveal nuances in strategy, design philosophies are revealed either in the artifacts themselves or, if not available in the object, then in the sequence of procedures used to manifest the artifact.

Richard's use of a folding chair, for example, reveals his commitment to simplicity. That this simplicity can be found in an object that can also

Richard single-handedly flipping a hull inside his shop. His practiced ability is a function of knowing the exact combination of lifting the hull to a balancing point and then how to walk the hull from one side of the bay to the other so that it can be lowered again.

Sometimes "shop tricks" are nothing more than the clever use of everyday objects. Here a folding chair conveniently holds the angle aluminum stock used as runners on the bottom of a hull at exactly the correct angle so that the runners can get welded in place.

accommodate a person is a nontrivial dimension to his use: here the chair is used as a fixture, holding up a piece of angle stock; later it will be used as a piece of furniture, holding up a person.

With the three pieces of aluminum angle stock in place but still sticking uselessly up in the air, it's time to bend them into place to become the skids that will protect the boat's bottom. Richard runs an acetylene torch up and down the angles, slowly dropping each one down until it lies flat along the length of the boat. He then welds them into place. While the boat is still upside down, he also welds some additional strips of aluminum bar stock along the outside edge of the boat, running forward from the transom about five feet. This, too, is for wear protection, as half-inch-thick bars are more easily replaced than sheets. (Other builders add even more wear protection in the form of high-density plastic skids or easily replaced wood skids set in a channel.)

Flipping the boat back over, Richard places the lid to what is now a bow deck box into place, welding the hinges to the front of the boat. While at the front of the boat, he also attaches a thick piece of aluminum that will become the padeye on the boat. At the back of the boat, he welds the first of two sets of braces that will help to transfer the power from the drive unit to the hull. This shorter set runs from the transom down to a wide, thick bar attached to the bottom of the boat. Because all these pieces are aluminum, they can all be welded together.

Later, when the upper braces are added, because they are made of steel, they can be welded to the drive unit but will have to be bolted to aluminum assemblies, made up of angles and bars, that are able to be welded to the hull. The junction is critical to the success of the boat. It is where power is transferred from the drive unit to the hull. But all the drive units, of all the builders, are made of steel and the hulls of aluminum. Aluminum and steel can be welded to themselves, but not to each other, and so each maker must confront the conundrum of power transfer across this unweldable divide. Each maker has determined a solution that, he feels, gets the job done based on his individual design philosophies. For some, it is elegance of solution. For others, it is durability. For Richard, the answer is simplicity.

Richard is not yet done reinforcing the rear of the boat; he adds triangular gussets in the corners and he also adds a strip that runs the width of the boat that will, later, line up with the leading edge of the steel bench where the driver and the engine will sit. He also cuts a hole at the back of the right-hand side and welds a short piece of aluminum pipe into the hole. Along with a plug that he will provide, this is the boat's drain. On the left-hand side, he welds in first a rectangular frame made of aluminum angles and then a U-shaped strap six inches above it. Together, the two assemblies, along with gussets to support the lower assembly, form the holder for the boat's battery, which will start the engine.

At this point, the bulk of the aluminum work is done. It is now time to work in steel. Richard gets right to it, placing two pieces of angle iron onto the transom. Although both pieces are the same length, one piece is narrower in profile, allowing it to nest inside the other angle. With the larger angle going over the top of the transom—which, remember, is a little over an inch thick at this point, being composed of both a piece of tread plate and a piece of aluminum channel—the narrower angle goes on the inside. In cross section, the two look like two upside-down Ls, and with their front edges aligned, they hold onto the two aluminum pieces rather like a plumber's wrench. Richard first tacks and then welds the angle irons together.

That done, it's time to roll out his steel worktable and begin on the drive unit. The heavy table accomplishes two things at the same time: it's a sturdy work surface not only capable of withstanding the abuse metalworkers mete out but also, with a ground clipped to it, carrying the electricity that makes it possible to weld. That means all a fabricator has to do is have reasonably clean surfaces; he can then stack metal pieces a variety of ways and have sufficient current to weld. This fact becomes critical when it is necessary to work quickly and efficiently.

Richard's first task is to build the base for the trunnion, which will attach to the assembly made up of the pieces of angle iron he has fitted to the top of the boat's transom. Like any hinge, there are two parts to the trunnion: the fixed piece and the mobile piece. He begins work on the mobile piece, which will eventually reach out to the drive unit. Onto a piece of half-inch steel plate already cut out with a half-circle arch on one end, he welds a cylinder. To get the perpendicular angle he desires—the trunnion needs to be plumb with the vertical line of the boat—he actually builds the assembly upside down, with the heavy steel plate resting on some scrap pieces of aluminum that are the same height as the cylinder. (Although aluminum can't be welded to steel, it can carry the necessary current to make welding

Richard building the pivot for the drive arm of one of his boats.

possible.) After double-checking that everything is square, he welds the cylinder that will hold the trunnion pin to the top, mobile, plate.

From these two pieces, everything else follows. Richard grabs two pieces of steel that have already been drilled and bent to hold the horizontal pin and pipe that will allow the drive arm to be lowered and raised; the vertical cylinder assembly itself will swivel the arm side to side. The forty-five-degree bend allows these two pieces to gracefully merge into the cylinder, and he makes a few additional cuts with the torch to make sure everything will come out neatly and then grinds the cuts that will be exposed to human contact later.

With the upper half of the trunnion made, it's time to assemble the lower half, which consists of only two pieces: a lower steel plate and the polished steel vertical rod that will sit inside the cylinder on the upper plate. The lower steel plate comes from a stack of plates on the floor, just as the upper steel plate did. Given the size of his shop and in consideration of how best to apply his own labor, Richard, like all the boat makers, has an inventory question to answer: which parts should he have made, which parts should he make ahead of time, and which parts should he make as part of any particular job? In the case of parts that he has made, a boat-builder has decided that he either doesn't have the equipment to do the job, the time to do it, or the interest in doing it. This can vary from part to part and maker to maker. In the case of the two stacks of steel plates lying on the floor underneath the plate roller, Richard has decided that the task of making the plates, with their two clean circular cuts on the inside and the half-circle arch on the outside, is simply something he would rather have someone else do. He orders a number at a time and stores them in a location where they can reside in inventory for several years, if need be.

Sometimes the demand for crawfish boats is sufficiently light enough so that Richard, like other boat makers, has to carry inventory over. None of them wants to, of course. Anything sitting in a shop or storage location is taking up space and not making money. On the other hand, there is nothing worse than a job you can't take because you can't make the part, or get the part, within the time frame that the customer requires. It's a difficult juggling act, not unlike what any small business faces, but the complexities of demand here are quite staggering. The demand for crawfish boats is a function of how many acres of crawfish are in production, which itself is a function of the price of crawfish versus other crops that could be grown on those acres, as well as the multilayered puzzle that is the agricultural supply chain and its regulations. A small change to the regulations governing rice or soybean production in the farm bill can tip the balance in favor of expanding crawfish acreage or curtailing it. The same can be said for changes in input costs: changes in the price of oil not only affect fuel costs, always a factor when you are running any kind of equipment, but also echo through the variety of chemical inputs on which farmers rely.

The complexities are rather overwhelming, and the builders themselves don't so much navigate them as wait for the farmers to reach a decision. Although farmers have to decide for themselves, and for the land they farm, what is the best course of action, they are not alone as they mull over their options. As each of the factors affecting the commitment of acres to producing a particular crop accrues, farmers find themselves gathering on the steps after church, in feed and seed stores, in equipment sheds, and even in the builders' shops to weigh options and get a sense of what others are doing. A consensus emerges, which can lead either to a dearth of orders for new boats or a welter of wants, such that builders have to promise boats on a first-come, first-served basis and often have to give purchasers, who almost always want their boats yesterday, the news that their boat will be ready weeks, and sometimes months, from now.

Such a boom-or-bust cycle is not new to Richard, who has weathered enough seasons of boatbuilding to have an intuitive sense of how much stock to keep on hand. Exactly how he reaches his decision is not something even he can fully articulate, but he frequents many of the same places as the farmers for whom he makes things and so the diverse streams of decision making are as open to him as anybody else.

With the bottom plate and shaft welded together, Richard assembles the two pieces of the trunnion and tests that everything works as it should. If everything is as it should be, he turns the assembly upside down and fits the angle-iron assembly from the back of the boat on top, blocking everything into the proper position with scraps of rectangular aluminum tubing of various sizes. The shaft that extends out of the bottom plate assembly fits into a notch in the combined angle irons. Everything is welded into position, including a small gusset made out of thick steel plate that is attached to the cylinder and the lower steel plate. It is not welded in during the earlier assembling because its position on the cylinder is somewhat relative to the position of the overall assembly on the angle irons.

Once the trunnion and its transom mount are together, Richard adds a pair of steel tangs that will allow a hydraulic piston to turn the boat. One tang is mounted directly in front of the trunnion and the other at the right end of the transom mount. He shims the other tang with a piece of steel bar so that it is fairly level with its mate. It's not clear just how much this has to do with keeping the force of the piston working only in one plane and how much of this is really a matter of his own aesthetic sense, which each maker possesses. Each of them can, and will, offer up a robust explanation for why they do something a particular way, and each detail does reveal something about how they imagine the machines they make and the forces that they endure and exert when at work in the world, but having listened to so many explanations, and asked—probably more times than any of the makers ever wanted to hear—"why did you do that?," there is part of me that believes that some decisions are less about function and more about form. That is, I don't know whether Richard has ever had a boat come back with a turning ram that misbehaved because its two tangs were not level, but for him, the leveling of the two parts is part of making the boat "right."

The whole assembly can now be fitted onto the transom, but Richard is not ready to bolt it into place. He will first assemble all the other steel pieces of the boat, and then paint them all at once before placing each one in its rightful place.

Richard grabs a J-bar, a three-quarter by ten-inch piece of steel with three right-angle bends already made. At one end of the J-bar is a circular cutout that will eventually hold the hydraulic motor attached to the wheel. The J-bar's bends are such that the wheel can be successfully mounted along the centerline of the boat, making it possible to have a steering-neutral point for the craft.

Although the J-bar is precut and bent, it does not yet have the bolt holes for the hydraulic motor, and the very first thing Richard does is to place the motor, mark the spots, indent them with a punch, and then drill them in a press. He then sets the J-bar on the worktable and blocks up a rectangular steel tube next to it. The tube already has the pipe for the vertical trunnion welded into place at its other end. Everything here has to be level and plumb, so he takes his time to shim and block things until everything is correctly positioned in the abstract three-dimensional space of the table.

This ability of the builders to think three dimensionally is something that needs more consideration than can be given here, but it's worth pointing out now, as we follow Richard's process, how readily the tables and floors are used as bases for building complex objects. The builders trust their floors and they have made sure that they are level. Tables are similarly constructed with great care so that they too will be level. With level foundations, makers can perform amazing tasks with little more than a carpenter's level, which also indicates when something is plumb, and a speed square, when something other than level and/or plumb is at stake.

It's important to remember that anything level or plumb needs to be checked in at least two directions—usually at right angles to each other. One could imagine an abstract plane, but a piece of steel plate works even better: such a plate could be level along its length but not level across its width. The hard part about such work is that, especially during dry fitting, which Richard has to do with the tubing and the J-bar, every action in one dimension can have a deleterious effect on the other dimension. Better builders anticipate complex outcomes, and they also have gotten very good at knowing to tack in one plane and then adjust in the other plane. In Richard's case, he quickly tacks the two large pieces together and then reinforces the assembly both vertically and horizontally with a number of gussets.

His next step is to build a second set of supporting braces, this time out of steel. He makes the braces out of steel tubing, and they are attached to the trunnion at the back of the boat through another piece of steel tubing that rises a foot above the assembly. Because the assembly itself pivots at the stern of the boat, a steel tang leads forward to a bolt that rises out of the intersection of the two braces.

With the braces in place a foot above the trunnion, Richard also has a place to attach the hydraulic ram, which will connect to the drive unit's arm. He welds a tang to the back of the vertical steel tube that rises from the trunnion, and then he welds a complex collection of pieces onto the arm, the net effect of which is to construct an elbow to which the cylinder will be attached. The elbow can be pinned into place or it can be allowed to float. The elbow has at its base a pin that passes through two pieces of heavy steel bar that are welded to either side of the arm's tube. Making it possible to keep it in place are two pieces of steel attached to a steel plate that is, in turn, welded onto the drive unit's arm.

Once all these steel pieces are fitted and welded together, Richard pulls them off the boat and paints them. (His signature color is a light blue-gray.) He also paints the drive wheel and the steel bench that will be included in the next part of the process. Everything is allowed to dry overnight.

In the meantime, Richard attaches the front wheels to the boat, using a piece of aluminum pipe that passes through two holes he cuts. Everything is welded into place and the pipe is reinforced with gussets on both the inside and outside of the hull. The aluminum pipe is sized to receive the steel shafts of the tires, which are bolted onto each side independently—the completed assembly gives the illusion of the two tires being on one shaft, but the aluminum pipe running through the boat is there for strength, not to contain an axle. The wheels are the kind found on small utility trailers throughout the region, and thus are something with which every farmer is already familiar.

The next step, often the next day during the week it takes Richard to build a boat, is to place all the steel pieces in their proper configuration. He begins by bolting the trunnion assembly first onto the transom, where six bolts pass through the two pieces of iron angle and thus through the aluminum tread plate and channel of the hull. Next, the braces are bolted into the aluminum blocks on the deck of the boat. With the trunnion in place, the drive arm is slipped and pinned into place. The elbow is added, and then the last major step in the assembly of the drive arm is to attach first the hydraulic motor that turns the wheel and then the wheel itself. Richard's wheels are fairly large, and while constructed of thinner steel than the arm, the wheels are still fairly heavy. He has an interesting arrangement for his cleats, alternating full-width cleats with narrow ones, which he has added to smooth the road ride of the boat. His cleats also have a steel rod welded to the back of the top edge, which not only decreases the wear on the cleat itself but also decreases wear on pond bottoms. With the wheel and hydraulic motor locked together through a bushing and key combination, Richard finishes this stage by attaching the splatter shield to the front edge of the J-bar.

With the drive unit in place, it's time to begin putting together the

Richard bolting the drive unit's steel braces to the aluminum hull. Because steel and aluminum cannot be welded to each other, each maker has to arrive at his own method for marrying the two metals together. Note the two tangs for the turning ram welded facing each other.

Richard preparing to run the hoses that power the drive wheel's hydraulic motor.

bench, which will eventually house the boat's engine, hydraulic pump, valve assembly, and driver's seat. The bench is a piece of sheet steel with edges bent down to give it rigidity. Where the bench attaches to the gunnel rail, Richard has reinforced it with some angle iron. A single bolt passes through each end of the bench and secures it to the gunnels.

Richard unboxes a Honda engine and places it on the bench. He then unboxes a hydraulic pump and bolts it to the engine. This combination of engine driving a pump will be the boat's single source of energy, not only turning the great wheel (forward and backward) but also powering the two hydraulic pistons. He positions the engine-pump assembly so that the pump's intake, which has been elbowed to face down, sits directly over a hole cut in the bench. He then positions the oil reservoir, a cylinder he makes himself, just ahead of the pump, so that the reservoir's output, a pipe at the bottom that faces forward, is in line with the pump's intake. The reservoir is held in place with two small flanges that bolt to the bench. The reservoir and pump are connected by a short length of hose that forms a U as it passes from one to the other. The hose is not under pressure and so it can be attached simply with hose clamps.

Next, Richard mounts the valve assembly that sits between the operator's seat, which is not yet in place, and the engine. As I have noted before, each builder has his own design philosophy, and Richard's desire for visual simplicity shows itself well here: he constructs a kind of manifold out of the three valves, each of which controls the three hydraulically powered parts of the boat: the two cylinders and the motor. The two valves that control the cylinders are spring-loaded, so that when not pressed in one direction or another, they return to a neutral position. The valve that controls the wheel has three settings, in effect—forward, neutral, backward—and is designed to stay in that position until changed by the operator. After bolting the valve assembly to the bench, Richard runs a hose, this time with pressure fittings, to the oil filter, which hangs off the reservoir by a pipe at the top of the tank.

Before continuing with the hydraulic system, Richard takes a moment to set a battery into its holder and to connect it to the engine. This seemingly sideways step actually makes sense when you realize that once the turning cylinder, with its attendant hoses, is in place, working with the battery becomes considerably more difficult. Richard's practiced economy is so fluid during assembly that you can almost miss simple gestures like this.

Each maker has his own usually hand-built oil reservoir
as well as his own arrangement of valves.

Richard takes especial pride in the tidy arrangement of the valves that control the flow of hydraulic fluid and thus the boat itself. Here, all three valves can be seen: the one in front steers the boat, the one in the middle lifts or drops the boat, and the one in back controls the boat's speed. In each case, when the oil has done its job, it returns to the reservoir to await being pumped and is pressurized again. The pump is attached directly to, and is thus directly behind, the engine seen here.

A mostly finished boat leaving Richard's shop. He will add the frame for the cover and the sorting table once it is outside.

The turning cylinder can now be pinned into place, and so can the lifting cylinder. The hoses that control them are run from the valves to each of them as well as to the hydraulic motor. Richard tests the entire system for leaks with pressurized air, and when he establishes that there are none, he fills the reservoir with 303 hydraulic fluid, a kind used widely in tractor hydraulic systems. He cranks up the engine and pressure tests all the lines, looking for any sign of leakage. With everything checked out, he neatens up all the hose runs, either zip-tying hoses together or clamping them into place where he has fittings to do so.

The boat can now be driven out of the shop under its own power. Once out in the yard, Richard adds any and all options that a customer may have requested. One option is a top, which is a collection of aluminum pipes that are held aloft with more pipes or square aluminum tubing and can be covered by various kinds of awning or sheeting materials. The tops are great for keeping off the sun, which is more of a factor late in the season, but they are even better at keeping the person in the boat dry from the rains that are an inevitable part of late winters and springs in Louisiana.

Another option is a sorting table, where traps can be dumped and the catch sorted and pushed into chutes where mesh bags hang. The trays can be simple flat-bottomed affairs or they can be made with a grate of aluminum tubing that allows small crawfish to automatically fall, via a large chute built underneath the grate, right back into the pond with no further action from the operator involved. I have seen some sorting tables made by Richard and others with removable grates so that the size of the bycatch crawfish can be determined with a great deal of precision. I have also seen pieces of garden hose, with slits down their length, pressed onto the grates to close up the distances. A final option for some of Richard's boats, but an increasingly popular one, is to have a piece of tread plate create a slightly raised deck between the forward deck box and the tire axle. It's very nice looking, reduces the trip hazard of the axle by itself, provides a place to stack crawfish sacks, and probably also adds a bit to the structural integrity of the boat. With all or some of those options in place, Richard parks the boat to one side of the drive in front of his shop. Like any new vehicle, he places a selection of useful parts and the manuals for the engine in the deck box so that when the boat's owner arrives to pick it up, the boat and its operator are ready to go.

If viewed from the rear and below, the equipment that allows the crawfish boat to be raised above the ground for smooth travel down field roads and sometimes highways can be clearly seen.

A completed boat awaiting pick up outside Mike Richard's shop.

THINKING THINGS

Understanding how land gets made is the first key to understanding how the crawfish boat got made. The same imagination that sees a topography beneath the surface topography—that sees, in some ways, topography at all—is the same, or at least intimately linked, with the imagination that first tacked together the odd collection of parts and pieces that produced the first boats.

But as we have seen, the entities and their relationships that make up this landscape are not the ones we anticipate. A landscape that is physically continuous, and hardly varied from an outsider's point of view, is understood by the people who live and work on it to be broken into pieces, each with its own history and particularities.[60] Discrete objects like men and machines that seem obviously distinct ignore our assumptions not only about the relationship between the two, but also about any possible alienation from work, the land itself, that the machine might impose upon the man. What we find, instead, is an intimacy that we usually only glimpse in paeans to artisans.

The anxiety about "machines in the garden" is not a new one, as Leo Marx chronicles in his history of American romantic pastoralism.[61] On the other end of the rocker arm that drives the discursive engine that dominates the American imaginary of material culture is American techno-triumphalism. The latter is, of course, the dominant one in most public discourses. It seems only natural, one supposes, that, in an attempt to offer either a critique of industrialism or an alternative, most humanistic interdictions would focus on those objects and practices that offer the greatest contrast. The resulting black-and-white, chiaroscuro image, however, loses all the subtleties that grays provide.

Neither the pastoral avoidance of technology nor the triumphal full embrace of technology are very interesting. Everyone has examples of both in their own lives; in the context of this book, perhaps the best example is the contrast between the individuals who cling to their old-feature phones, proclaiming that all they need to do is make a phone call, and the individuals who have to have the latest smartphone, with whatever-sized screen that covers the entire side of their face when they use it. Neither is wrong, of course, and neither is right; there is a huge middle ground for everyday users of phones who would, yes, like to make phone calls, but also check the weather or get directions. Once upon a time, as it were, this would require several devices; now it only requires one.[62]

Drawn into conversation with either one of these individuals, we find ourselves quickly at a discursive dead end; there is no arguing with someone who insists they only need a phone, just as there is no arguing with someone who insists that some particular gizmo, or feature of a gizmo, has radically changed the way they live. But neither conversation is particularly invigorating, neither advances any ideas except those already held by their proponents.

Contemporary humanistic discourse about technology, especially in my field of folklore studies, sometimes feels much the same to me. There seems to be this toggle between the premodern and the postmodern, between people plowing with horses and the virtual reality of FarmVille. What's left out, of course, are the practicing farmers who actually populate the landscape, trundling across fields in tractors with, yes, enclosed cabs that perhaps to the untrained eye make them seem more like a lunar rover. But the image of the sunburned farmer smelling the earth as it is turned over must be counterbalanced against the danger of breathing in soybean fuzz or rice chaff or dust, against the danger of working in the rain, of being exposed to hornets and wasps, of pressing on when it really is too hot or too cold to be working but the work must still get done.

Unseen but no less present alongside the farmer in the field are the tool and equipment fabricators and repairmen. Men like the Olingers, or any of the others listed here (and many who are not), have a kind of virtual presence in the fields, precisely because they are in constant dialogue with that landscape through their work for farmers. All it takes is for one farmer to walk into a shop with a problem or a broken tool. The problem or the nature of the break first gets defined through a series of gestures and explanations. Eventually, a potential solution emerges, perhaps from the man in the shop or from the farmer; perhaps the men develop the idea together, or perhaps the problem gets left in the shop and a later conversation with a different farmer makes the solution apparent. No matter. The idea gets made or applied to the break.

But at this point it is still only an idea. It remains an idea when it leaves the shop, despite being manifest in metal. It is a tool when, and only when, it is applied and it works. Failures are described as seeming like "good ideas at the time."

This may be the extent of the innovation. Or it may be that another farmer passes by the field and sees the new tool in action, or he hears the first farmer talking about it after church, at the mill, or at some other place or event. Or maybe that other farmer arrives at a shop with the same, or similar, problem and the people inside present him with this possible solution. Suddenly or slowly, depending upon the moment, a one-off, individual idea becomes something larger, a communal idea. If it sticks around for a while, as the crawfish boat has, and it gains a kind of historical depth, including variation within a larger stable form, then we have something that folklorists call *traditional*.

Unfortunately, it seems like this tradition of innovation, which is just one of what must be many—I have glimpsed a half-dozen more, some small and simple and some quite complex, in my short time studying the boat—is largely left out of most humanistic treatments of the agricultural landscape. And, again, the agricultural landscape is only one among many landscapes, or however else we want to name such complex cultural ecosystems, that escape our notice because we too often focus on the nature of the materials and not the nature of the imagination.[63]

We need not look far to find parallels to the economy at work here: it is exactly the one to be found in the early twentieth-century wheelwright shop of George Sturt, a schoolteacher who, upon the death of his father, suddenly found himself in charge of a shop much like those we have glimpsed here. His initial response to the work, in particular to the milieu in which he found himself, was quite poor; he wanted little to do with the customers who seemed forever wanting something and rarely willing to pay what it cost, let alone for the time—the most precious resource of all—that they consumed. Fortunately for us, and for Sturt himself, he eventually settled into his new discipline and began to appreciate the many men who came through his shop, both farmers and workers, for who they were and what they brought to the life of the shop and the community it served.

One thing Sturt came to appreciate was the wisdom to be found in

A crawfish rover, or, as it is known affectionately at the Olinger shop,
Fred, built on the vision of a particular farmer.

the complex interweave of the men who made things, the men who used things, and the things themselves: "The nature of this knowledge should be noted. It was set out in no book. It was not scientific. . . . The lore was a tangled network of country prejudices, whose reasons were known in some respects here, in others there, and so on. In farm-yard, in tap-room, at market, the details were discussed over and over again; they were gathered together for remembrance in village workshop; carters, smiths, farmers, wheel-makers, in thousands handed on each his own little bit of understanding, passing it to his son or to the wheelwright of the day, linking up centuries. But for the most part the details were but dimly understood; the whole body of knowledge was a mystery, a piece of folk knowledge, residing in the folk collectively, but never wholly in any individual."[64]

While we might today quibble with Sturt's assertion that the body of knowledge was not scientific and that it was, in some fashion, a mystery, his assessment that what was known was distributed across a collection of individuals strikes me today as very astute. One of the singularly great advances in folklore studies has been the focus on what particular people know and how they communicate it to—or in the case of certain verbal forms, "perform" for—others. What this has allowed us to do is not to presume any particular collection of individuals is in possession of some indiscriminate bolus of knowledge.[65] Rather, tradition is differentially known, and called upon, within a group, with each member a node within the larger, living network that is itself always, like a net, unraveling and being repaired, and even at times made better.

What worried Sturt, as it did, and does, many others, is what happens when this seemingly fragile system is changed as radically as occurs when the technological innovations of the Industrial Revolution are introduced and change occurs at what seems like an unprecedented pace. Suddenly, economies that are centuries old, economies that structure social relations, are swept away. Some cried good riddance, while others, like Sturt, worried that such change would disturb too much too quickly and that, given a little time to accommodate and determine their own future, most likely "local needs were exchanged for cosmopolitan wishes."[66]

The icons for this change, for Sturt, were "steam and steel." Steel, in particular, seems to stand in for the industrial age in a way that hammered and forged iron does not. Unlike its ancestor, steel arrives in shops already in a variety of useful shapes extruded in some distant factory foundry, shipped great distances, and delivered on the back of flatbed trailers. It can be cut only with very powerful tools, and in order for it to be assembled, a fairly large electrical pulse is used to melt two pieces of steel into a common joint. Steel, by its very nature, calls forth an entire global economy where complex systems of production intertwine in ways hard to glean at a glance, one where the individual always seems insignificant and ineffective.

Steel and its equally malleable cousin plastic make up almost every instrument we regard as modern—cars, cell phones, computers. Both are prized by a wide range of industries for their plasticity and so we also think of them as the stuff stamped out by factories and that, in the process, stamp the life out of the men and women who man the machines, and who are unmanned by them. But like Chaplin's *Modern Times,* such a view is largely a projection of our own ennui and anxieties onto others. These others, were we to stop them on the shop floor, might very well reveal themselves to be fully in control of their humanity by not only having more say than we think imaginable in the making of ordinary things that actually make modernity work but also—perhaps on occasion, perhaps regularly—enjoy repetitive tasks that require the use of their hands as well as their minds, and indeed find our work—stuck as we are in offices stacked with paper, straining to read under fluorescent lights and often having to answer to a nebulous group of individuals known as administrators—quite debilitating.

This is, in fact, a regular response in any conversation I have with farmers and fabricators when we talk about kinds of work. While they envy my air-conditioned office on summer days, they often state that they just couldn't do what I do, and by that they don't mean write books or teach classes. They mean sit in an air-conditioned office all day and not move very much. For them, to think is to move. For farmers, it is to move around and across a landscape, constantly assessing what needs to be done and what can be done in the time one has. For fabricators, moving around means moving oneself in volumetric space in order to construct three-dimensional objects with the kind of precision that will allow them to withstand the abuse heaped upon them. It is this moving while thinking, moving as thinking, thinking with movement, that seems so important in trying to understand the imagination that lies at the heart of this landscape and behind the crawfish boat itself.

The nature of that movement is, as folklorist Richard Bauman is fond of saying, to be discovered. In the case of farmers water leveling, like Dwayne Gossen, riding high up in an eight-wheeled tractor, despite the power they wield, they are, in their own minds, more like men crawling on the ground, eyes closed, feeling their way. When he rides in the enclosed cab of an eight-wheeled, articulated tractor, one imagines that the operator is, if not quite a disembodied mind dully driving this way and that, then at least so alienated from the interaction between machine and landscape as to rely mostly on visual cues and the scant few sounds that make it past the roar of the engine and the insulation of the glass windows. Nothing could be further from the truth. Having ridden extensively both in these giant tractors while farmers plowed as well as in combines while they harvested rice or soybeans, I can safely attest to how little they actually pay attention to any and all gauges and readouts that report engine RPM, grain flow, or the height of grain in a hopper. Instead, farmers are constantly "feeling" and "listening" to the machines in which they ensconce themselves in order to get work done.

A farmer's sense of a tractor is a complex set of cues that, taken together, give him a great deal of information.[67] There are, of course, the gauges that most farmers use only when getting used to a new piece of equipment or to double-check their own sense of what is happening when they feel a machine is reaching its load capacity. But, really, the gauges remain simply that, a check to what is sensed quite literally; the man in the machine is not simply driving it by pushing levers and turning the wheel this way and that while alternating his glance from out one of the windows to a panel of gauges and back. He is doing those things, but while he is doing those things he is not only listening with his ears for sounds the tractor makes either in direct response to the work, or something going wrong with the work, but also to indirect responses the machine has to a particular task; the sounds an engine makes when it is idling easily versus beginning to take on a load versus straining under a load are very different, even though the RPMs on the relevant gauge may appear the same.

Along with those sounds come things that are, more than anything else, felt in the body. They range from very large events, like the tractor crossing a levee, to middling events, like feeling the clunk of a drive shaft engaging or a hydraulic cylinder reaching the end of its stroke or the tractor encountering an unintended (and perhaps unseen) high or low point, to the small vibrations that communicate the texture of the ground or the working status of the machine.

On the larger end of things, the sensations are perhaps obvious even to passersby: a tractor, or a boat, angles up over a levee and down again. Or perhaps an operator needs to mow grass in a ditch or work along a slope. In all these cases, his sense of balance, his sense of the center of gravity, is rooted in the aggregate of himself and the machine. Mature operators know how far their equipment can safely tilt or lean; apprentices are the usual cause of damage, which is why they are frequently put on smaller, older gear as they slowly develop the larger kinesthetic sensibility that efficient and safe operation of farm equipment requires.

In the middle range of what we might call instrumental kinesthesia are what farmers feel as the ground passes underneath them unseen while water leveling or perhaps not seen in great detail while high up in the cab of a tractor. Previous plowings or the running of a crawfish boat can often result in ruts being created in a field. From the farmer's perspective, these ruts are undesirable when growing rice, since they can mean low spots where water may get trapped or they may, if they are long enough, drain the field inappropriately. In either instance, the ruts disrupt the farmer's ability to control the water level in a field with the kind of granularity preferred. These ruts are felt as small, sudden drops in the body of the tractor, and their width is gauged by the jolt upward that follows. Most farmers have a very acute sense of the speed of their vehicle and thus typically a fairly good idea of the distance traveled between two moments in time. (It also helps that they have had this ability to gauge distances and dimensions reinforced by knowing the width of a rut created by a crawfish-boat wheel or by another kind of plow; these two kinds of information, one visual, but in memory, and one kinesthetic in the present, are combined while water leveling to afford them a high degree of precision.) Depending on the depth of the rut and the overall fit of the tractor, there may be a concomitant sound made by the tractor, which might also be felt. A tractor with a somewhat loose fitting somewhere, for example, will make a distinctive clunk, which many farmers will listen for, often knowing that the clunk is only prompted by changes in depth of a certain size or kind. The sound, caused as it is by a movement within the tractor and not of the tractor in relationship to the landscape, may also be accompanied by a second cue.

The finest range of sensation are almost proprioceptive in nature, felt as vibrations of the body, and occasionally accompanied by a sonic cue of some kind. These sensations are produced by the tractor's engine and reveal to the operator the degree to which the engine is under a load. Farmers typically describe this as feeling or hearing the engine strain, and it is, I confess, one of the more nuanced types of perception that I have come across in my years of research; there is little to no obvious change in the pitch or the volume of sound these large diesel engines make. At three hundred fifty horsepower or better, the engines in these tractors are capable of pulling a water plow through the water with relative ease, and it is not unusual for them to be doing so at extremely slow speeds. Because the plows can push so much water in front of them, farmers must work at slow speeds in order to make certain that they do not slop water over the small levees that outline the cut. Water loss is less of a concern than topsoil loss. Thus, the larger engines are run at what almost seems an idle, heard and felt as a low rumbling. As the plow being pulled picks up water and mud, however, the engine begins to work a bit harder, and farmers listen and feel for that moment when, perhaps, the engine will need to be fed a bit more fuel.

In all these instances, farmers are highly attuned to their tractors. They describe this process in two different stages. The first stage occurs when a farmer is just starting off, just learning how to farm and how to work with equipment. As a teenager working with an older family member or friend, typically fathers and sons but sometimes uncles and nephews, a farmer has to learn to "feel the seat," as one young farmer told me. It is a matter of learning how to feel the bottom of a field with the tractors' tires, the young man noted, and in doing so, he reached out and down with his arms and spread out his fingers, as if he were imagining himself crawling through the water, feeling with his hands to determine how the land lay.

A farmer learns these things on a particular piece of equipment, and so the second stage occurs when he transitions from one piece of equipment to another, because each piece of equipment has its own feel, not only as a piece of machinery but also as a sensing device. Another farmer, Kip Link, who had recently purchased a John Deere tractor after using nothing but Case tractors for twenty years, noted that it was going to take a great deal of getting used to: "[A John Deere] tractor runs different, works different." The same observation occurs when a farmer has gotten used to the feel of a particular brand of equipment and that manufacturer makes a significant change to the drivetrain, the suspension, or some other facet of the machine that requires the farmer to "recalibrate" their senses.

〰〰〰〰〰

It is this distributed cognition—this blend of mind, body, and machine—that is so striking, that seems so fundamental, precisely because it reveals how intimate the relationship is between humans, technology, and the landscape. What makes it especially interesting is how clearly everything can be discerned. This clarity is a function of the fact that the work of leveling a rice field is a discrete task in which the criterion of a successful performance is clear: the field drains evenly, with no low spots or high spots. Not only is it the case that a farmer depends on his skill to level a field; it is also the case that his skill is subject to a wider, often quite objective evaluation. Internally, a farmer who does not know how to "feel the seat" spends too much time either looking at gauges in front of him or looking behind him to double-check the state of his plow. Looking at those things distracts him from seeing where he is going, resulting in inefficient plowing or in very slow going. Externally, any field can be, and usually is, observed by other farmers. Uneven fields reveal themselves by patches of rice that are a different color or a different height. Such mottled fields raise questions and comments in nearby equipment sheds, in agricultural supply stores, and after church. Its opposite, a field uniform in color and height, receives appreciative nods and comments: "That's a nice field of rice you've got out near the highway."

Such an outcome depends on a profound knowledge of the topography of the landscape, including a sense of the underlying geology, as well as a highly attuned sense of one's equipment. Mediating the relationship between the equipment and the landscape is a function of a collection of abilities and sensitivities, which someone like the anthropologist Charles Frake would term a high order of "cognitive ability" (1985, 255) that makes it possible for a farmer to know, when he is leveling a field, for example, how much dirt he is moving and from what unseen hill it is departing and to what unseen hole it is arriving.

In being a performance with clear objectives, a lot of farm and fabrication work gives us an alternative view into the nature of the human mind. Most studies of cognition take place within controlled environments,

〰〰〰〰〰〰〰

such as in university laboratories, where thousands of undergraduates' responses to various kinds of stimuli have come to dominate what we know about how we think and what that reveals about the nature of the human brain. Plucked out of the flow of life, out of the world itself in many ways, laboratory experiments allow scientists to isolate a task, removing possibly complicating and confusing variables. Such empirical study is best when it is balanced by the kind of rich, multivariate work that the ethnographic study of human behavior provides. Recognizing the need for real-world tasks that take place within a constraining context, psychologists have had a historical fascination with navigation at sea, particularly navigation by Micronesian sailors traveling between islands that typically lie many days' sailing apart, because the task of navigation is clearly defined and has easily measured goals—notably successfully arriving at one's destination. Navigation, in other words, provides us with a nice display of cognitive performance.

The idea of performance as a clearly marked field of action is a familiar one to folklorists, who embraced the ethnographic study of defined human behaviors half a century ago. As folklorists discovered, understanding the parameters of a given performance is as important as the performance itself. In the case of cultural performances with multiple individuals, sometimes from multiple groups, the parameters are multidimensional and often quite difficult to sort into orders of importance and influence. In the case of something like navigating at sea, like the plowing of a field, the parameters are fewer and it has been easier to elicit the kinds of cognitive work accomplished. "The lesson to be drawn from these studies [of Micronesian navigation]," Frake observed, "is that the islanders' seafaring exploits do not depend on some uncanny intuitive powers, nor on personality quirks driving people to seek danger, nor on the luck of lost sailors adrift at sea, nor even on rote-learned 'local knowledge.' Instead these navigational abilities depend on a profound general knowledge of the sea, the sky and the wind; on a superb understanding of the principles of boat-building and sailing; and on cognitive devices—all in the head—for recording and processing vast quantities of ever-changing information."[68]

In his own work, Frake extends the investigation into displays of high orders of cognitive ability by examining the navigational work of medieval sailors, who developed a rather robust system for correlating lunar with solar time in order to be able to predict the tides of the ports on which they called. Such information was, in most instances, critical to determining when one could enter or exit a port, and what route one would need to take in order to do so. The medieval sailor was thus able to do something that modern sailors, dependent on first tide tables and now computational devices, cannot determine for themselves with a high degree of precision: the stages of tidal activity.[69]

Why concern ourselves with medieval sailors? Precisely because they relied so heavily on their compasses. The compass is not, as Frake points out, "a time-finding instrument [but] a very abstract model, a cognitive scheme, of the relations of direction to time, of solar time to lunar time, and of time to tide."[70] As such, it is correlational thinking embodied in an artifact, the product of the human mind setting out to accomplish a complex task that also has very clear outcomes. Such accomplishments open up a space for investigation of where the processes of sensing and thinking actually lie. They are neither completely inside us nor in the tools we use but rather seem spread across both.

This way of thinking about the relationship between the thinking we do and the things with which we think is described by Edwin Hutchins as "distributed cognition." An anthropologist, Hutchins has sought to bridge the gap between his own field and psychology, between culture and cognition as objects of study. Conventionally, of course, the two are considered distinct areas of inquiry, but only, as Hutchins observes, because the boundary between inside and outside have been so firmly drawn, which "creates the impression that individual minds operate in isolation and encourages us to mistake the properties of complex sociocultural systems for the properties of individual minds" (1995, 355). Hutchins's argument is that cognitive sciences have overallocated intelligence to the inside of human subjectivity. The problem with such a view, he believes, is that it mistakes, potentially, one dimension of a larger system for the system itself.

Perhaps the best example of mistaking the mind for the larger system within which it operates, Hutchins notes, is the famous "Chinese Room" thought experiment created by John Searle. In an attempt to argue against the possibility of artificial intelligence (a case of technology thinking on its own), Searle set out the following scenario: he is locked in a room where messages in Chinese are slid under a locked door. He himself has no knowledge of Chinese, but he does have a book that allows him to determine the character sequences and to respond with a correct sequence of characters that he then slips back under the locked door. The outside observer perceives a meaningful reaction, but, given Searle's role in the communicative instance, was there really any meaning? Searle's response is no, of course not, and thus something like the Turing test (a test designed by computer scientist and philosopher Alan Turing to determine a machine's ability to exhibit intelligent behavior indistinguishable from that of a human) is misguided at best.

But if we set aside the intent of the argument, which was to demonstrate that "syntax is not sufficient to produce semantics," something really rather interesting pops out at us: what Searle has done is encapsulate what might be described as a sociocultural cognitive system. On his own, our experimenter cannot communicate from inside the locked room, but as an ensemble, he and the resources inside the room can. That is, "the cognitive properties of the person in the room are not the same as the cognitive properties of the room as a whole."[71] The attribution to an individual mind of an entire system effects a kind of surgery in which interaction and our chief means of interacting, our bodies, are removed.

Reduced so, the unhooking of cognition from interaction becomes clearly absurd. What we need is to study cognition more as it occurs in the world and study cognition less as a limited set of responses from an individual isolated in a laboratory. Hutchins proposes we call such work *cognitive ethnography.* Returning to some of the language used by Frake in his description of cognitive psychology experiments, I am struck by the occurrence of the word *performance*—not just the use of the word but that it is used in ways folklorists would easily recognize:

> We are concerned here not with judgements about the mentality of
> an age or the wisdom of a culture, but with the cognitive abilities of
> individual human beings. For evidence we must turn away from assess-
> ments of the strangeness of a culture's beliefs or the weirdness of its
> symbols to an examination of performances that can be seen as displays
> of cognitive ability. But what counts as such a performance? Probably
> most things a human being does should count. The problem for the
> investigator, and sometimes for the performers themselves, is to know

what the performance is. "What's happening?" Or, in psychologists'
language: "What is the definition of the task?" (1985, 255)

Psychologists, of course, prefer to define their own tasks and remain
anxious about user-defined tasks as being vulnerable to collusion. Folklorists and others who are used to working from the outside in see this less as
a vulnerability and more as a matter of openness. Such an openness to the
"task world" allows us to form different understandings of what people do
with their minds, with how they think and with what they think. As we
have seen, thinking must be broadly understood.

TOWARD A NEW UNDERSTANDING OF CREATIVITY

There are more makers of boats than are mentioned here. The ability to make useful things out of metal is diffused across the landscape, though perhaps not evenly. Since embarking upon this project, I have entertained many a call, note, or conversation from a colleague or acquaintance who in talking with me has learned more about crawfish boats and fabricators than they ever wanted to know and finds that he or she has suddenly developed an eye for blue sparks and boats with wheels. I have trekked into woods thick with mosquitoes and poison ivy or squelched into the mud along rice fields to examine a lot of boats. Some of those boats sitting deep in the woods or in the "long grass," as they say around here, are earlier models from one of the makers mentioned previously. Some are clearly someone's attempt, some more successful than others, to make a boat for themselves.

We can never know what motivated such individual attempts. It's not uncharacteristic of farmers, at least in Louisiana, to be known for being very attentive to the money they spend, and so some are no doubt motivated to try making a boat out of a desire to save money. But there are others who enjoy trying their hand at making one because they are good at welding, have a design or novelty they want to try, or want a break from farming. All of these are reasons to try making a boat. And the many different kinds of boats that I have glimpsed through thickets of grass or ivy are testaments to the widely available set of skills that many individuals in south Louisiana first encounter in agricultural shop in area high schools. (Indeed, Harold Benoit remembers working with students on drawings for boats as a way to expose them to drafting.)

But as the uses of the boats became more demanding—the drive units became more powerful and hulls became stronger and larger to withstand the crossings—the range of makers actively involved in the production of boats slowly constricted to a few individuals who possessed not only the skill set but also the facilities and the economic wherewithal to make the new, substantially improved form of the boat. The modern hydraulic boat, as it is sometimes called, is a far cry from the first boats, which were not much more than two commercially available items, a boat and a tiller, cobbled together. Such a transformation from an assembly of widely available parts that were as easily had as the closest Sears to a custom product requiring knowledge of two kinds of metalwork, aluminum and steel, as well as an understanding of gear ratios and power distribution in hydraulic systems, meant that production that was once fairly evenly distributed across the landscape became centralized to a few nodes.

This particular form of distribution is not new, nor anything to be lamented; it represents a fairly old form of development of an industry, one dependent on a variety of abilities focused on a particular artifact. Once upon a time we called the intertwining of skills and an object *craft*. The men who possessed the skills that led to particular kinds of products were known as *craftsmen*.

Craft, as ability and its application, is our topic here, and it is, it should be noted, a popular one—a resurgent one, it would seem. Folklorists have long taken craft as one of their central subjects, since craft is usually dependent on a dense network or, really, set of networks that intertwine people, ideas, and practices. That is, craft foregrounds the role of culture in our lives. A craft is typically imagined as a set of skills organized around a particular product, be it a tangible object like a boat or a quilt or an intangible one like a legend or a song. In most instances, those skills are probably generally diffused across a group: lots of people sew or a good number can weld, and plenty of people tell jokes and countless others hum a tune while they do something else or to comfort themselves or a child.

Occasionally, however, someone has an especial affinity or ability for a task, or at least he applies himself sufficiently that he comes in some way to be noted, marked, for being able to do a particular thing. Not everyone is asked to bake a dessert for a family gathering, and not everyone is consulted for advice about cars or computers. Rather, there are a handful of people within any particular family or community to whom expertise has been attributed. To them, we bring our problems. From them, we anticipate solutions—or at least a very nice slice of pie. These are the artists and craftsmen in our worlds. Too often we transmute their focus and their willingness to practice into something like a talent, a gift given from an ethereal realm as opposed to a hard-won ability that was perhaps, yes, *driven* by an as-yet-to-be-understood curiosity or desire.

There may very well be something to someone "having a knack" for hearing a melody—I need look no further than my own daughter's seemingly perfect pitch as opposed to my tin ear—or that others understand the world through movement in space as opposed to words—Isadora Duncan is often quoted as protesting, "No, I can't explain the dance to you; if I could say it, I wouldn't have to dance it!"—but our understanding of this internal, cognitive landscape is in its infancy. The mapping has begun, but no matter how well it might one day explain the origins of ability, it cannot detail the development, for that lies in the externalities of the physical world of the individual and the social world of practitioners.

The folklorist Henry Glassie has charted this dynamic landscape in describing the development of folk housing in Virginia, of folk song in rural New York, and of folk narrative in Ireland. Perhaps nowhere does he describe the process more precisely than in articulating the way Turkish women assemble carpets of intricate design by doing nothing more than knotting together strings. The designs, of course, exist nowhere but in their heads, but how do the designs get there, and how does a woman know how to manifest that design through what amounts to a pixelated drawing, row by row, of a larger scheme? There is no sketch over which paints are applied. There is no plan lying on the floor by her side, occasionally drawing her attention. Instead, "a weaver is alone in concentration and part of a team at work" (Glassie 1999, 51). Glassie observed that each act of a weaver "collects the whole of her biography" because she often grows up playing in the shadow of her mother weaving, first collecting scraps of yarn, getting a feel for the fiber and the colors involved, then making her first few knots and learning first the techniques and then the designs, and finally making a carpet of her own while her mother, family, and friends look on (52–53). Craftsmanship, in many ways, refracts larger questions about human nature, of humans as individuals and as part of a larger group. As Glassie mused while watching the Turkish weavers, "To be human is to be alone and not alone, at once an individual and a member of society" (51).

Our farmers and fabricators who went on to become boat makers are

Dale Olinger welding a drive wheel for a crawfish boat.

no different. They are drawn to it because it represents an intellectual or technical challenge and/or because it represents an economic opportunity. The acumen they bring draws the attention of others. Many want a boat, but lack the ability or the desire to make a boat for themselves. As we have seen, the initial experimentation was fairly diffuse, with, as Ambrose Olinger once remarked about people making "all kinds of jackleg contraptions." But refinement quickly concentrated around a handful of makers, some of whom are still making boats while others have moved on to other pursuits. Even the reasons for getting out of the business of building boats are diverse: for some, interest in building boats waned; for others, it was no longer as profitable as it once was. For at least one, retirement called, the next stage in a life already richly lived.

This is a terribly important point that cannot be made often enough: any community seen from the outside, from an initial glance, appears homogeneous to the untrained, unfamiliar eye. One equipment shed on a farm or one welding shop at the end of a gravel drive looks like another when seen from a car traveling too fast on a country road. But the similarity ends there, with the superficial, with the snapshot, with the windshield. Once you are out of the car and into the shed or shop and you have spent some time watching people work and listening to them talk, the differences become apparent. Personality and experience always matter and they are always different.

～～～～～～

Our current moment imagines everything as networks. Perhaps that is as it should be. Although there is no denying the trendiness of the term, and perhaps even the conceptual frameworks behind it on occasion, the idea of a *network* is not out of place here. When I have referred to individuals as nodes with differential abilities (read characteristics or properties) distributed across a landscape, I have in fact been invoking a network. A more properly objective study might try to determine a delimited number of characteristics—perhaps skills established as important to and by the community involved—and then assign a numerical value to each, indicating how capable a given individual was at that skill as determined by independent observation as well as the assessment of the community involved. (This isn't as hard as it sounds;

everyone knows who is capable, who is great, and who is unlucky
. . . a lot.)

The projection of various attributes of the nodes of this network, as an abstraction, could reveal a number of interesting dynamics, and perhaps one day I will do that work. For now, what interests me is the diversity and diffuseness of the network as a living thing, as the people themselves move to and fro on the landscape, into and out of shops, into and out of churches, into and out of fields. As they do so, ideas shuttle among them, passing from one to another. Sometimes the idea is only a fleeting thing, a comment made about a particular moment. Sometimes it is a suggestion, or a request.

One such moment occurred one summer afternoon in the Olinger shop. Gerard was in the middle of making a series of PTO ditchers for various farmers, one of whom had come to the shop to check on the progress. Farmers will often drop by ahead of any stated deadline in the firm belief that shop men spend too much time talking and need to be reminded that people are waiting for equipment to be fabricated or repaired. Most of their visits are taken up, of course, with talking with the men in the shop; the talk confirms the farmer's suspicions and the shop men shake their heads about how farmers have so much time to waste. This particular farmer, Jimbo Hunley, gazed at his half-completed ditcher and mentioned that while he admired its design in general, it would be nice if he had some way to check the chain without taking the covers off.

"Could I get a hole?" he asked.

Gerard thought about it, and eventually arrived at a solution that did not add complexity to the build or weaken the ditcher in any fashion. PTO ditchers spin at a high RPM in order to scoop dirt with their blunt steel blades and throw it thirty to forty feet away. The power is transferred by one or two heavy chain belts. The overall assembly has to withstand both the torque of the tractor's PTO as well as the grind of being dragged through the dirt behind the tractor's right rear wheel. The result was a two-inch hole cut into the ditcher's steel tube that was threaded so that a pipe cap would close it up—the same kind of cap you might see on a sewage clean-out on the side of your house.

The members of the Olinger shop dubbed it the "Jimbo hole" that year. When other farmers came to pick up their ditchers, they asked about the new development, and it was not long before all ditchers coming out of the Olinger shop were equipped with Jimbo holes, or Jimbos as they came to be called. Several years later, the innovation has largely become so much a part of the form that it has largely lost its name.

Similar innovations occurred, as we have seen, in the crawfish boat: at first, there is a lot of wild variation, with lots of cross talk among makers, often in a very confused fashion, but then there slowly emerges a preferred solution whose adoption ripples across the strands of the network. In the case of the crawfish boat, the first significant novelty to be widely adopted was the hydraulic drive unit. The second was the locally made hull.[72] And the third was the addition of the front wheels. Each moment in the history of the form enabled the next.[73]

This series of seemingly discrete steps suggests to observers outside the actual network a kind of logic, or narrative, of progression. But to understand progress as linear is to focus on the form and not on the network of people making the form happen. Progress is not a straight line, but rather more like a pinball seemingly bouncing from node to node at random. (Perhaps there is some randomness built into such systems, but that is something, alas, yet to be explored and documented.)

The experience of such progress is not, as the idea of randomness might suggest, chaotic. Rather, each boatbuilder works within his space, within his node if you like, and ideas flow through. But rarely do ideas flow unencumbered; ideas are almost always accompanied by a variety of metadata, verbal and material commentary on the nature and quality of the idea. Verbally, farmers and equipment operators report on successes and failures using a piece of gear, or they express a desire for a feature to be added or for a particular feature to be removed or modified to be made more, or actually, useful. They will also comment on things they have seen on gear belonging to others. Farmers will report a conversation with another farmer about a boat by another maker; that is, farmers report on their own comparison of notes. Shop men listen, nod, and process all of this according to their interest and their mood.

In addition to the listening they do, and sometimes at the same time, they are looking over the equipment that has come in, looking for stress fractures, failed welds, or places where metal has worn smooth or through. They are trying to understand what has failed, since all mechanical things wear and fail, and whether it has failed because it was worked too hard or worked incorrectly. When the piece of gear is not of their making, they are especially keen to understand how the other maker has anticipated such things, and whether they have built in any extra rigor that is perhaps concealing the gear being used hard or incorrectly and for which they must accommodate in any repair or revision they make. They are also looking for good ideas. They don't necessarily need to be in possession of bad ideas to desire good ones; there is, in a fabrication shop, always room for more good ideas, which can be sorted and stacked for future use much like the templates and jigs that hang on most shop walls.

And so ideas flow through each shop in multiple modes, and their capture occurs in much the same way they always have: abstracted sufficiently and held in the mind or fairly quickly adapted and put into production. In the case of adaptation, the capture is rather straightforward and easily observed: the kind of hydraulic valve found on another boat appears on the next boat produced, or, when sufficient stocks dictate a delay in adaptation, as soon as it is time to place the next order. Valves, engines, rams, motors, pumps, welds, bends, curves, lengths, widths: all are available for adaptation.

But sometimes the adaptation is not immediate; sometimes the maker wants a little more time to process a new idea. It may come as a surprise that, even though cameras are available on almost any device we carry, I have never seen a single maker photograph something. It is just as rare for them to take notes. Instead, what I have seen again and again is a process in which something is abstracted to the point where it can readily be mentally filed away. If the idea is, for example, a wider hull, a maker will survey the rest of the craft to see what other changes were made, and whether they appear to be original to the new design or themselves later adaptations as the operator discovered the need for increased power or better, or worse, handling characteristics. Is the worn-on waterline the same as on other boats, or is it higher or lower? Handling characteristics deduced, mechanical advantages and disadvantages tallied, the maker reduces the change in hull size to something like an *if-then* statement: "if the width increases, then the power needs to be increased," or, more likely, something like a *for* loop: "as I increase the width, these things result." Such a compression allows makers to stow the information they need away and to draw upon it again when the moment calls for it. Anything larger or more detailed would simply require too much time.

So ideas flow readily throughout this diffuse network. Innovations travel from node to node either by example or by word of mouth. Sometimes the solution to one problem actually solves another. One of the common complaints about the rear wheels is that they create trenches in the fields, especially when the person fishing the field tends to run the same route each time through a field. The ruts will be left when the field is drained and it can be difficult to smooth them with a plow the following spring when it comes time to plant rice. (I have ridden in articulated tractors with eight wheels six feet or more in diameter and felt the ruts.) This has largely seemed an intractable (pardon the pun) problem with various solutions proffered—Olinger had gone to two six-inch-wide wheels set two feet apart as one solution. Then, about three years ago, Kurt Venable began to weld steel rods onto the edges of his wheels' cleats. The problem he was trying to solve was how quickly a piece of three-eighths-inch thick steel four inches long can get worn down to a nub, sometimes, depending upon the composition of a farmer's soil, in a single season. It turns out, however, that the reinforced cleats not only create a wear bar but they also ride a little better on field bottoms and dig a little less. This was, all the builders agree, an unexpected bonus.

~~~~~~~~

Like any folklore form, there is a great deal that all the boats have in common as well as room for variation—in many instances, that variation is only apparent to someone who has looked at a lot of crawfish boats or who has had a maker point something out. Sometimes the differences are manifest in the form, and sometimes they are in the manufacturing process that lies behind the finished form. When I first began approaching the boat makers to ask them about their work, I admit that I worried that they would be reluctant to talk about their processes. But in an era in which protection of intellectual property is the norm, none of these makers was concerned, for the most part, with anything being secret. Sure, each has a manufacturing trick or two that he thinks gives him an edge, but each man is fairly certain that he builds the best boat. Just as important, they all have worked on each other's boats or have at least seen them. And, as I have already noted, farmers are not only a source of feedback and ideas, which are variously received, but they are also a conduit for information about developments by other builders. (More than one maker made a sideways comment about the "Farmer News Network.")

Given the size of the market, and its apparent porousness where almost everything that can be known is known, as well as the money at stake, one could easily imagine that the urge to mark out a territory would be overwhelming. Surely some sort of patent regime would have emerged by now. Surprisingly, none has, and it is not because these men know nothing about the legal and/or institutional ramifications of not filing a patent. Nothing could be further from reality. Each of them is surrounded by the legal necessities of owning and operating their own businesses, of needing lines of credit extended from suppliers and to their customers, and of needing insurance not only for themselves and their employees but also for anything that might happen in and around shops filled with high-powered equipment capable of bending, cutting, or melting thick pieces of steel and aluminum.

There is, in fact, a patent granted for a "Crawfish Harvesting Boat" (patent number 4817553), which is described as:

A multipurpose flat bottom boat for harvesting crawfish. The boat includes a deck on each end with an open spaced hull between the decks. The boat is propelled by rotatable wheels which may have a propelling units on their ends and are mounted on telescopically extensible members.

The rest of the patent goes on to describe a boat with nets on either side—one assumes a kind of seining as if one were fishing for shrimp—and trays to receive the crawfish as they are periodically dumped out of the nets. From the trays, the crawfish are then cleaned, sorted by size, and dropped into waiting containers, eliminating the need for further processing.[74] From looking at the accompanying drawings, it is not quite clear how all the mechanics required to make all those things happen would quite fit on a small boat, and I don't know that the boat's inventor, a Russell Knott of Arnaudville, ever built a prototype.

The only other patent on a crawfish boat I was able to discover using the United States Patent and Trademark Office's official search process was one for a toy boat, which is somehow fitting. For one, it is not clear that anyone could claim to have invented the crawfish boat—although that does not seem to stop people from patenting things in other arenas—and anyone who did would face a fair amount of "prior art" (as examples of things preexisting any claim of invention are called).

No one ever came forward to claim the idea, and not one of the makers, when asked, expressed any interest in patents. Kurt Venable's response was particularly telling, because he in fact does have patents for the Cajun microwave that he used to manufacture: "Oh, I thought about it, but I decided it just wasn't worth the trouble. My other patent didn't do me a whole lot of good."

Unfortunately for our present moment, we seem captivated with patents as a means of measuring creativity. It has helped to create the sense that creativity has shifted from the independent inventor, the subject of the so-called golden era of invention said to have lasted from the late nineteenth century to the early twentieth century. The lone individual, tinkering in his or her garage or basement, still exists, but he or she is usually imagined as coming up with something we might see on late-night television infomercials or home-shopping channels. If they make it into the mainstream, it is because they invented a toy or game that can be cheaply made. Anything with a larger impact is going to require economies of manufacturing and economies of thinking that just aren't available anymore. (The Internet momentarily shifted this perception for a short time, but even now small inventors seem mostly to get absorbed by preexisting large companies that are seen as better able to refine or support the innovation.)

But perhaps the problem is not the independent inventor but that we require invention to be independent. What the inventors, refiners, makers, and users of the crawfish boat establish is that innovation can be diffused across a network, a network made up of very different individuals with very different experiences, desires, and abilities. We can spend time talking about each of the individual nodes. We could even find some way to measure their contribution, but how would we measure Bill Krielow's being so insistent that Harold Benoit should make him a boat that he wrote him a check sight unseen? How would we measure the practically innumerable comments that goaded each of the makers to change a little something here, and a little something there, and in the process resulted in the robust, mature, hydraulic crawfish boat that you can pass on a highway every spring in the Louisiana prairies?

Measuring discrete facets is tantalizing, but it misses the larger point of how the whole operates as a system that eventuated in an amphibious agricultural vehicle produced in sufficient numbers not only to make a living for its makers, but also because it contributes substantially, and occasionally solely, to the livings of its operators. Three decades ago there was no such thing as a crawfish farmer, someone who only grew and harvested crawfish. Now there is. That is an amazing accomplishment. And those responsible for it are all around us.

BACKWARDS

The present and its close cousin the recent past are always inevitably moving beyond the reach of memory into the narrow, dark confines of the distant past. If we only ever attend to the latter, we lose the opportunity to broaden it. The distant past will remain narrow, and our reason for being nostalgic, for all the things we have lost, will linger. By letting our past narrow, we also limit our present and, concomitantly, our future. That is the heart of the matter, the crux that lies at the center of folklore studies: living tradition is a dynamic entity. Distilled from the variety of practices within which it lives, it becomes a different kind of thing, subject to different forces and available as a different kind of resource, the kind of resource that others might use for purposes far from, and sometimes at cross-purposes with, the tradition from which it originated.

A fine example of this came to me one spring, which is always a pleasant time in Louisiana. A good portion of the population is yard and garden mad, and there is a certain joy that permeates the season, as people try to do as much as possible before the long haul of the region's hot summers comes to dominate the days. Perhaps due to the area's being culturally Catholic, Lent's variability calms things like the proverbial oil on water, and the social calendar isn't quite as full as it is in the autumn, when it would seem there isn't a weekend that goes by without some town hosting a festival.

But oil on water—that is, actual oil on actual water—seemed to be the problem of the day at the wooden boat festival in Saint Martinville. A crowd of people had turned out to stroll among the boats on display along the bank of the Bayou Teche at what was billed as an event that was both a "Wooden Boat Congrès [Congress]" and a "Re-enactment of the Arrival of the Acadians." At one end of about a football-field-sized stretch of the bank, sitting in the shade of a massive live oak, was a pirogue, held aloft on sawhorses. At the other end was a cluster of larger boats, one of the most impressive of which was a twenty-five-foot aluminum boat decked out in Coast Guard colors with two massive outboard engines hanging off its stern. Stretched in between, in a kind of replication of a historical timeline, were a few more pirogues and a half-dozen wooden boats with one-cylinder engines that are affectionately known locally as *putt-putts*, named for the sound their engines make. There were a few kayaks and canoes, but they laid scattered on the ground, in the sun, at the far end of the line, and clustered with the big, modern boats on trailers, suggesting that they were at the wrong end of the timeline, the Johnny-come-latelies of the boat world. In a word, too modern.

Normally this day in the middle of March is the Acadian Memorial Heritage Festival, but during the past few years the boats have garnered more and more of the attention. In years past, the festival has stretched from the Acadian Memorial building itself to its sister museum, the African American Museum (together they are part of the St. Martinville Cultural Heritage Center), turned the corner at the Evangeline oak tree and worked its way up a good deal of the two-block boulevard that stretches from the bayou to Main Street. In 2013, gone were most of the crafts booths, replaced by food vendors, and much of the action of the festival had shifted to underneath the oak itself and the alley of trees that has been maintained as a park.

Perhaps things were as they should be. History is such an intangible thing, more akin to a ghost than anything else in our lives. The boats had appeared at some point in the previous decade of the festival's occurrences and they had proved an immediate favorite. There had always been a diverse collection, but under the umbrella of the Evangeline oak, the pirogue emerged as history materialized, so much so that in the previous year the Louisiana state legislature had declared it the "Official Boat of Louisiana."[75] The Acadian Memorial responded by proclaiming 2013 to be the "Year of the Cajun Wooden Pirogue."[76]

None of this surprises. We live in an era where jeans arrive already faded and sometimes with ripped hems on store shelves, where furniture in window displays and catalog photos already have their edges rounded from use. We live, in short, in an era longing for the timeworn. We buy things and assemble them in carefully curated collections in our homes. They are there to nurture us when the press of the present threatens to overwhelm us. They are necessary fictions, fictions that ground us, fictions that offer us something like authenticity.

And what could be more authentic than a "Cajun Wooden Pirogue"?

It's just that the closer you get to the compound, the more things seem to fall apart. The word *pirogue* is borrowed from the Spanish *piragua*. The original Spanish term refers to a wide variety of dugouts, and, in fact, the Acadian Memorial website includes the "dugout canoe" as one of the boats gathered under the auspices of the Wooden Boat Congress. The earliest pirogues were, without a doubt, dugouts, and we even have a documentary film, in color no less, of Ebdon Allemon making the pirogue that would become a central prop in Robert Flaherty's *Louisiana Story*.[77] (And what could be more idyllic than a young boy, paddling freely about a landscape he knows well with his pet as a companion, which, in this case, is a raccoon. An exotic landscape with exotic inhabitants using an exotic vehicle.) Dugouts are fairly universal forms of boats, however. There is nothing particularly Cajun about them, except that the Cajuns learned to make them from the Native Americans, and perhaps from the enslaved Africans after their arrival in Louisiana. The Cajun connection really represents the fact that the Cajuns were still making them, and using them, long after other groups had moved on to other forms.

Make no mistake: the pirogue, like any personal-sized craft, is an extremely useful vehicle. You can get away for an afternoon of fishing or hunting or just lazing about in a shady spot under a cypress tree. I have done all three. But wooden boats, unless they are finely finished, prefer to stay in the water. Out of the water, they crack. Getting them in and out of the water also requires a trailer because they are, compared to their fiberglass, plastic, or aluminum counterparts, quite heavy, which often also makes them a bit harder to maneuver in tight spots.

Nevertheless, they are made out of wood and nothing quite says "olden days" like wood—or the use of *olden* for that matter—and so stretched out beneath the oaks and cypress trees that line the bayou, we have our venerable Cajun wooden pirogue, an adopted craft with an adopted name. As I stood there and admired its positioning at the head of the line, with the powered wooden boats coming next and the massively powerful aluminum boats last, a video-production crew backed its truck up next to the aluminum boats and began to unload tangles of wires and boxes. What were they here to photograph?

They made a beeline, as best they could while juggling all their equipment, for the pirogue, only stopping when a festival organizer pointed out that their truck was blocking the street. They were there, it appeared, to capture not only the pirogue but its ceremonial christening with, as it turned out, a can of Diet Coke and the reenactment of the Cajuns coming to southwest Louisiana. Pirogue photographed, reenactment captured, the video crew packed up its gear and left as quickly as they had arrived on the scene. After all, if you want to tell the story of Cajuns in south Louisiana, or of boats in south Louisiana, how can you not tell the fuller story, the one that proceeds not up to the last of the wooden boats, but takes you all the way to the aluminum and steel boatbuilding industries that employ hundreds upon hundreds of skilled workers in high-paying jobs that don't come and go with the popularity of a guided reality-television program but are there every year, come rain or shine? In short, how did the video production crew miss Jimmy Gravois's boat? They had parked right next to it and unloaded their gear in its twenty-five-foot-long, fifteen-foot-high shadow.

No, they had seen the boat all right, but they hadn't *seen* it because its story was not worth telling. And the story is exactly the same story as that of the crawfish boats or the shallow-water boats: true stories about indigenous creativity that are happening in our own present and not some distant, mostly imagined, past. Jimmy Gravois's business began with small aluminum fishing boats. After twenty years of making fishing boats, the company decided to try its hand at a government contract, winning a bid to make small response boats for the Coast Guard. Today, that contract means Metal Shark, Gravois's company, must complete one boat a week, and each boat must meet the rigorous standards of the Coast Guard's inspectors, who are so particular about welds, for example, that men weld in parallel seams only a few inches apart, just so each weld is consistent, with no irregularities from a change in personnel, around the entire boat hull. Metal Shark employs hundreds of people to design, manufacture, and equip the boats they sell to organizations and companies from around the world. Like the handful of men who shaped the crawfish boat or the handful who shaped the shallow-drive boat, Metal Shark represents an economic force with considerable staying power that somehow remains invisible and thus unavailable for analysis in terms of tradition and creativity.

Metal Shark is quite literally making history—as were, and are, the farmers and fabricators who invented the crawfish boat. And as were, and are, the handful of companies who invented and refined the shallow-water drive-boat industry. They are making history, but somehow they remain outside of it. This confounds me, even now, as I think about it. How could this be?

One part of the answer, I think, lies with the fact that folklorists occasionally lose our focus, leading to a failure to communicate that the artifact is only an index of the mental process, and it is the process that is important, not the artifact.

Mental processes are, of course, notoriously hard to document, but folklore studies has long led the way by carefully documenting the focused output of such processes: the things people say and do repeatedly. Because we investigate not minds alone but minds in a network of cooperation and competition, inflected by an array of desires, experiences, and biases, we also understand that such processes and culture itself are highly interdependent. Things like boats manifest culture in a tangible way, but they are not alone in doing so, as Henry Glassie noted: "Buildings, like pots and poems, realize culture. Their designers rationalize their actions differently. Some say they design and build as they do because it is the ancient way of their people and place. Others claim that their practice correctly manifests the universally valid laws of science. But all of them create out of the smallness of their own experience" (1999, 227). That smallness, of course, includes other people, and as people move in and out of others' experiences, they communicate ideas through, and in, the forms available to them.

Culture is itself, as the anthropologist Fredrik Barth observed, "nothing but a way to describe human behavior" (1969, 9). In particular, we use it to describe repeated behaviors, especially those behaviors that differ from one group of individuals to another group. But that is all culture is, a collection of observations of things said and done on some sort of consistent basis in some sort of consistent form among a group of individuals, each of whom has his or her own experiences of the things themselves as well as having his or her own desires, which may converge or diverge from the group at a particular moment in time. Because of the complexities, folklorists and anthropologists have historically been drawn to groups of individuals who are more or less insulated in some fashion precisely in hopes of understanding the cultural processes of transmission, reception, and retransmission of ideas with the goal of documenting which ideas stick and which ideas occur only once. The nature of the insulation, which we occasionally have mistakenly idealized as isolation, has varied, but the goal has always been to reduce the number of variables involved in the transmission and reception process in order to get a better grip on the basic dynamics.

Close attention to instances of saying and/or doing and a slow building up of one instance upon another gives a consistency to the fabric not only of ideas held in common, and thus amounting, perhaps, to something like culture, but also a better understanding of ideas held alone; the momentary idea that goes no further is perhaps as interesting as any other, but it is difficult to document, unless one happens to be there to get it. It is also, as Petr Bogatyrev and Roman Jakobson pointed out many years ago, more difficult to assess as a product of a general human competence than the idea that gets repeated. The difficulty, for understanding creativity, is to be able to document not only the emergence of a new idea but also its acceptance into a larger community. The two actions taken together give us a better understanding of the creative process as it unfolds in the world.

Understanding the necessary tension between the two, the moment of the unique individual utterance and its admission into a larger set of communal ideas is a lot harder than it looks. This kind of hermeneutic approach—quite literally the establishing of a horizon within which an economy operates—is often ignored in the press so as to put a face to an invention, to make the phenomenon more dramatic. This kind of impatience results in the weird transformation of folk art into outsider art under the aegis of the art world where the relationship between idea creation and community reception is often so transmuted that audiences regularly find themselves confronted by the artifacts of individuals driven, by whatever means, to the edge of human understanding. Alternately, from the entertainment industry, we encounter in reality television increasingly strange, or estranged, individuals, a natural by-product of assuming that personality matters more because somehow it is more visible. Together, the two regimes, the art world and the entertainment industry, have done what Jorge Luis Borges once observed about the Russian novelists of the nineteenth and early twentieth centuries: they have proved no one is impossible.

Is folklore studies responsible for this madness? Where we have relied on extant notions of genius, of the particular, and thus sometimes peculiar, individual, we have participated in the cult of personality that can lead to these ends. In fact, as Glassie observed, a number of the individuals with whom we do our best work are often themselves somewhat estranged from their communities. Their somewhat estranged but still native perspective is often a natural bridge for us as we try to locate the center of a community and establish its networks of ideas and individuals.[78] Given our own ethic to champion the disempowered, it is only natural that we sometimes feel a deep affinity for such individuals. We are there, we quite often imagine, in an attempt to salvage the margins before they are lost to the steady advance of time. We are there, we hope, to get people, perhaps there or perhaps the audiences of books like this, to rethink what is central and what is marginal.

Salvage and advocacy are enterprises in which folklore studies historically has been involved, as Richard Bauman and Charles Briggs make clear in their history of the field, but that should not preclude us from writing about absolutely normal people doing absolutely normal things. When we look to the present, when we look to what not only remains economically viable but is actually an innovation as well, we find ourselves in the presence of a handful of men who have made or continue to make indigenous amphibious vehicles that are products of both an individual fancy and of a real economic need.

I remember the first time I showed up in Kurt Venable's shop. He heard me say I was from the university and he immediately led me into the part of the shop where his computer-controlled fabricating machines are. Numbers and percentages poured out of his mouth that I could not hope to follow. After a little bit, he paused, furrowed his eyebrows, and stood looking at me.

"They are very nice machines," I said.

"What do you want to know?"

"I don't know. I don't think anything. I'm not really interested in them."

"Why not?" he asked.

"They don't make the boats. You do."

Later, Venable explained that he had had a number of people visit his shop over the years, but they mostly came from engineering backgrounds and were interested in the machines—because we keep mistaking things for the process.

At the other end of the academy, the machines often get in the way for humanists, who conceive of power tools less as tools bent by individual will and more as indices of a homogenizing modernity against which we must struggle. If you think about it, this idea makes for an interesting complement to the engineering faculty's assumptions about the machines being the dominant player within the context of these small shops. Both sets of scholars think the machines change the nature of the work to such a degree that they do the work, that they are the work.

While it is true that Kurt Venable and Gerard Olinger, for example, use machines dependent on the electrical grid, made in Iowa or China or Germany, and forged of high-tech alloys to craft boats with aluminum hulls, steel drive units, and complex hydraulic rigging, they remain very much in control of the process. The machines are, albeit very powerful, extensions of their will and their imaginations. And it is living testimony to the strength of those imaginations that these men, like the others chronicled in this book, have carved out an economic niche for themselves that allows them to live well, send their children to school, and take vacations as they like. The powerful machines that hulk about in their shops are no more than chisels in the hands of a wood-carver or a lump of clay in the hands of a potter. No, you won't find them perched at the edge of doorways in a pool of sunlight surrounded by a dusting of wood shavings. Instead, they work their magic in metal often under the luminescence of mercury arc-vapor lamps or fluorescent bulbs, because the work they do needs to be protected from the wind or rain or it requires precise control over the lighting. Or, just as likely given the heat of Louisiana's summers, it's just nicer away from the door.

While some may regard these makers as too close to the large shadow of modernity, historians of technology overlook these makers as too small a concern, either in terms of impact or in terms of innovation. Perhaps worse than simply being small, some historians even consider their work derivative. David Edgerton, for example, coined the term *creole technology* as a way to avoid the "innovation-centric picture of technology" that dominates the history of technology, and yet his recasting of the relationship between vernacular practitioners and some larger, and hazily defined, entity like "modern industry" seems to hinge on the idea that those outside of centers of industry and power have "not been a significant technical innovator in recent centuries" (2007, 85). For Edgerton and others, those who build at the fringes of modernity are mostly interesting in terms of how they continue to use older technologies after the rest of the world has moved on. Edgerton's cases are the bicycle in China or the continued production of the Volkswagen Beetle in Mexico and Brazil into the seventies and eighties.

Such a view completely misses the role that small shops play in this environment, and I am sure in others as well. None of these makers is working with yesterday's technology. Rather, they are making the tools that farmers need to get the job done and that, as all of them will tell you, the centers of industry miss either because of the scale or because they simply can't see it. The Olingers are one of a number of shops that make the rear-mounted PTO ditchers that shoot dirt far from where it might otherwise block the edge of the ditch, and they are also the only shop in the area where you can get a side plow installed: each is custom fitted, and they are quite often put on tractors straight from the dealer. If hydraulically powered side plows and high-RPM ditchers are yesterday's technology, someone better alert the past, because farmers back then would have liked to have had tools as useful, well made, and powerful as these.

For many folklorists, as well as others interested in making, Edgerton's *creole technology* will seem remarkably like Claude Lévi-Strauss's idea of *bricolage*. Both concepts accept standard definitions of centers and margins, with the modern being the center and everything else being, well, merely what plays in the shadow of the smokestacks. Having spent time among these shops and the many shops that line Route 90 and serve the oil industry here in Louisiana, and catching glimpses into the contents of shops throughout the South and Midwest, I would argue that these images of center and margin are really a function of where researchers tend to be and not where the work actually happens. That is, such landscapes as the agricultural landscape have become unconventional for the study of innovation and production not because nothing happens there but because we just don't go there often enough.

How did we come to this impasse? How did we come to this moment where an entire sector of our economy, and of our culture, is not visible even to observers as supposedly sensitive as those found in universities? Certainly academic specialization has played a part, and within those specializations too often conventional wisdom prevails, but how do we undo what we have done? How do we make it so that vibrant parts of our world do not slip into institutional interstices? Not producing art, not practicing an established craft, not big enough economically or historically, not technological enough, these makers would seem to be homeless in the modern American academy, which is perhaps why folklorists are best suited to bring them to the attention of others. More important, they offer the discipline of folklore studies an effective counterbalance to being seen as antiquarian in disposition. There is nothing wrong with wanting to capture the old ways of doing things. But these makers, and others like them, are in no need of salvage, only careful documentation as they go about their daily lives of making a reasonable living doing what they love, fully engaging both their hands and their minds—always at the same time.

Were I in a position to reshape how things are organized, I think I would reinvigorate the old agricultural extension model, but I would give it the chance to be more dialogical in nature. As a practice, agricultural extensions mostly still use the center-margin model, wherein the agent traveled across his, and occasionally her, territory, bringing the latest developments in farming techniques and technologies to the residents. In some ways, it is pastoral vision. Better agents always saw the opportunities in conversation over a lecture, and in fact some of the earliest photographs of the development of the crawfish boat came from former extension agent Dwight Landreneau, who was kind enough to share his slide collection with me when I first began this project.

But what if we took this model and increased the amount of time agents spent collecting information about what people are doing and why (they think) they are doing it? It would be a practice of documentation that is quite familiar to folklorists, and the agents would also be in a position to deposit the materials into active archives so that the information could be examined by a wide range of scholars, who might find that this stream of data opens up entirely new vistas for research and knowledge creation. Of course, I am biased in believing my own field is well suited to do such work, because our training is to be patient, to wait for the larger idea and sometimes story to emerge; because, I hope, we wouldn't miss the big aluminum boat gleaming in the sun because we would not arrive with the story already told and only in need of some local color for fleshing it out.

Why does this matter? While there's nothing wrong with archaeologies of earlier American lives, we can't just keep people circumscribed to the past if we want them, and us as scholars, to have any sort of future. It's time Benjamin's angel of history was allowed to turn her head and see the present moment in all its beauty. Because the present is most poignant when it is understood as being about both the past and the future. Because we have to frame our work in such a way that we cannot take away from people in the present the ideas they have drawn from the past in order to create the future as they understand and wish it.

At some point in the future, perhaps even sooner than I can imagine, there will be men and women in these shops who no longer cut and bend and fuse aluminum and steel in order to wrestle out of bars and sheets things like boats and ditchers. Instead, perhaps, they will walk into their shops, spend their day tweaking a three-dimensional design using the knowledge in their heads that they have gained through their own successes and failures as well as the successes and failures of those around them, and then they will push a button and a machine of some kind—perhaps we will still call them printers—will begin to build through a careful series of overlaying traceries the desired object. Perhaps it will be steel or aluminum, or perhaps it will be some futuristic material like plasteel.

In that moment, however, do I want folklorists in the future to lament that no one runs a lathe anymore, shaving bits of steel at high RPMs that endanger the worker and result in waste material that has to be swept and canned and shipped back to a smelter, doubling the amount of time and distance and handling that stock material undergoes? I would rather future folklorists look past the technology, past material, to understand how the makers of their time discover places where their individual abilities and proclivities fit within a larger sociocultural matrix such that they make a decent living, contribute to the economic success of their community, and demonstrate the power of ideas when they have a proper audience. That's a future we can all live within.

ACKNOWLEDGMENTS

I would like to reverse the usual progression of acknowledgments, which often in scholastic texts such as this begin with one's academic mentors and peers, works through various other obligations, and ends with the family. Without a doubt, I am deeply embedded to any number of individuals who have trained me, urged me on, required more of myself than I thought myself capable, and buoyed me up when I fell into dismay, but more than anyone else the daily care and feeding of my thinking and writing must be credited to my wife Yung-Hsing Wu and my daughter Lily Wu-Laudun. That this book is done, that you are reading these acknowledgments, must be attributed to them.

My larger family is to blame for my love of things. I grew up in a household headed by an architect and an interior designer. I was expected to understand spatial volumes in a way that was never required of words. (And I am grateful for it.) I also grew up in a household of gadget lovers: my father and mother held a deep-seated belief that surely the solution to a problem was just around the next technological corner. Lucky is the child who is presented with such a parade of "things with great promise." If nothing else, the crawfish boat is a manifestation of that promise to the fullest degree.

Beyond the walls of my home and the well-worn paths to my family there lie the men and women who gave so freely of their time and of their lives. In many instances, they feel like extended family to me, no one more so than Gerard Olinger, who put up with countless dumb questions with such good humor and patience. No less patient was his brother Dale Olinger, who is always the first to call out my name when I walk into their shop. The larger Olinger family has shown me no end of kindness. Much the same can be said of Dwayne Gossen, who always let me ride with him. Anthony Kramer and Jerry Leonards were also great helps, answering questions with warm smiles.

The other makers interviewed for this book were just as helpful and just as kind: Kurt Venable challenged me to get things right, and I can only hope that this book meets that challenge. His wife Sheryl was always sure to make me feel welcome and the two of them always made room, and time, for me in the office at the front of their shop. Mike Richard allowed me to sit beside him and take notes and draw diagrams while he built a boat as well as tag along as he visited with various clients. I will always remember his slightly bemused smile as he got on with his work as I hung over his shoulder. There are many more, some of whom feature in these pages and some of whom are present in moments of understanding both large and small. (All confusions are my own.) I thank them all.

There are, in a work like this, always intellectual debts to pay. Some are obvious. My debt to Henry Glassie, for example, is clear. Henry's role as my mentor both as a student of his and later as a young, and then older, scholar is everywhere present in this book: in the topic, in the nature of the writing, in any number of small choices that will escape most readers, happily, but will, I hope remind him of our many conversations. Others who shaped me both as a scholar and as a person were Richard Bauman, whose precision was something I have long admired and hope to one day acquire; John Michael Vlach, whose gusto for life and for knowledge always cheered me; and Lee Haring, whose steadfast drive to make sense of things that matter has urged me on when I lagged. I never had a class with John or Lee, but they were my teachers all the same. I did have a class with Susan Gubar, and I once promised her I would never tell, so I won't.

I would also like to thank my colleagues in Louisiana, not only fellow folklorists Marcia Gaudet and Barry Ancelet, but also Clai Rice, who has been a stubborn interlocutor, and Brandon Barker, whose work in cognitive folkloristics ran parallel to my own, more philosophical approach. And I would be remiss if I didn't thank Carl Brasseaux: being the great man that he is, he will never recognize how much his friendship and collegiality mean to me.

Two final thanks: one to my very patient editor, Craig Gill, of the University Press of Mississippi, and one to Carrie Roy and the staff of the Louisiana State Board of Regents, who made it possible for me to have the time to get Craig the manuscript he kept reminding me about. I hope their efforts, and his wait, have proven worth it.

It should be noted that work for this essay was supported in part by an AT-LAS grant from the Louisiana State Board of Regents (LEQSF(2013–14)-RD-ATL-09).

THE MAKERS

The men and women discussed in this book who continue to operate businesses open to the public deserve your business, if you have the work. If you are just curious to see imaginative work still happening in the American landscape, then feel free to drop by, but, be forewarned, there is always work in need of doing.

Paul Abshire Welding Works
    301 E. Mill St.
    Kaplan, LA 70548
    337-643-7933
Hughes Welding and Manufacturing
    1300 Airport Rd.
    Jennings, LA 70546
    337-824-2176
Olinger Repair Service
    190 Olinger Ln.
    Rayne, LA 70578
    337-334-8099
Mike's Aluminum Welding
    7433 Eunice Iota Hwy.
    Eunice, LA 70535
    337-457-8664
Quirk's Welding and Fabrication
    3712 Hwy. 10
    Washington, LA 70589
    337-826-9353
Venable Fabricators
    901 West Branche St.
    Rayne, LA 70578
    337-334-2933

NOTES

1. Andrews (1922 [1999]), 2.

2. James (1982), 3–5.

3. Of course, from a somewhat larger perspective, Gothic cathedrals arose on a rather limited landscape and, as John James has documented, involved a limited set of artisans who traveled about Europe, from site to site. That is, any phenomenon looms large when seen from a particular perspective, obviously, and we would do better to be more open to zooming in and out, focusing more on the underlying dynamic and not always worried about the size of its impact, which has been one metric by which creativity has been measured.

4. In this work, I am heavily indebted to Henry Glassie, whose work in this regard is considerable. Early in my education, I was most influenced by his *Folk Housing in Middle Virginia* (1975) and, as I note a little later, his *Passing the Time in Ballymenone* (1982). But perhaps the work of his on which this book is most closely modeled is his *Turkish Traditional Art Today* (1993), although clearly the present volume misses the portraits of makers contained in that work. Glassie's own work in the latter volume is reminiscent of Charles Zug's *Turners and Burners* (1986), which was also an influence in my own development in material culture studies. In addition, John Michael Vlach's *Charleston Blacksmith* (1981) is exceptional in its focus simultaneously on individual makers and the cultural and social systems within which they work. His study of Philip Simmons, an African American blacksmith working in Charleston, is especially illuminating in this regard.

5. See especially Comeaux, Elias, and Dasgupta's "Creativity, Cognition, and the Case Study Method" in *Frontiers in Cognitive Psychology* (2006), but work by Dasgupta in general.

6. Glassie (1982), 15. *Passing the Time in Ballymenone* is an ethnographic study of an Irish community that reveals that there is plenty of middle ground not only in the nature of our existence, but also in terms of how we pursue our path through that existence; there are folks within a community who stand out more than others while remaining firmly a part of the dense web of community relations and ideas. *Ballymenone* is filled with thousands of performances, some verbal and some mate-

rial, all of which are conscious manifestations of seemingly simple country folk living their lives year by year. Glassie came to adopt the local term for such individuals: *stars*. He notes: "The star stands at the center. Any consideration of a work of
art, a story or song, in Ballymenone leads you to an exceptional individual. . . . The
District's culture is not something apart from the particular individuals who are
the force of its coherence, the reason for its existence" (681). For Glassie, the older
Irish men telling stories about saints in the past who journeyed across the same
landscape that the men did in a present filled with bombs and bullets revealed that
it is people who get along in a world filled with others who may or may not be to
their liking.

7. See Glassie (1982), Zug (1986), and Gilmore (1999).

8. In the arena of surface-drive technology, it's how Warren Coco and K. P.
Provost got their start. It's a well-established dynamic and probably in many ways
reflects the older guild system.

9. The observation is part of Glassie's introduction to his ethnography of a small
Northern Ireland community on the border with the Republic of Ireland. His con-
tention in that study is that the easy solutions of sociologists and politicians pay no
heed to the stories that populate a difficult landscape.

10. Ancelet has given a number of papers at meetings of the American Folklore
Society on the topic of how residents of south Louisiana in general and Cajuns in
particular have been represented in various forms of media. He has even taught
a course or two at the University of Louisiana at Lafayette. One of those courses,
titled "Imaging the Cajun," paired documentary films, of varying quality and accu-
racy, with the studio films they quite often inspired: for example, Robert Flaherty's
Louisiana Story (1948) and Anthony Mann's *Thunder Bay* (1953). (Please note that
the pseudodocumentary nature of *Louisiana Story* has been the subject of considerable debate.)

11. In fact, as their studies continued, it became increasingly clear that the
Cajun and Creole communities had long been not only socially intertwined but also
culturally intertwined, leading Ancelet and Brasseaux, among others, to steadily
expand the scope of their research, although they stayed largely within the purview
of Francophone groups. (Brasseaux's multivolume history is especially compelling
in this regard [1987, 1992].) To be fair, this had been their initial charge when they
began their work and so they mostly sought to complete that work, encouraging
others to pursue the many possibilities that their work had created.

12. Rosan Jordan and Frank de Caro (1996) take up the matter of how the
consistent imagining of Louisiana as a "folklore land" is, in fact, a deferred classism

that some early writers embraced as a way to effect their own social mobility. It is, as their essay suggests, not unlike colonial/postcolonial efforts and effects found elsewhere in the world.

13. Such occurrences are not unknown in the region. The first oil well in the area was drilled near Jennings in 1901. In the years that followed, a lot of wells were drilled. Both during the boom time and as production slowly played out and wells were removed, a lot of pipe, concrete, and other materials were left behind in fields. Farmers regularly get their plows dulled on bits of chain and other industrial debris.

14. George Sturt's *The Wheelwright's Shop* (1923).

15. There are a number of excellent studies by folklorists of individual craftsmen or artists, like John Michael Vlach's *Charleston Blacksmith* (1981) or Douglas Harper's *Working Knowledge* (1987). There are also a number of studies that focus on a category of craftsmen, like Charles Zug's *Turners and Burners* (1986) or Janet Gilmore's *The World of the Oregon Fishboat* (1999).

16. This is a far different approach to the question of technology, especially in agriculture, than what Martin Heidegger worried about toward the end of his life. Heidegger worried that the world was engulfed in a "technological frenzy," one in which we imagine that technology will solve fundamentally philosophical and/or ethical dilemmas. For Heidegger, to defer human dilemmas to technology was to lead a life filled with dread: technology's very objectivity could just as easily solve the human dilemma by simply getting rid of humans.

17. Richard Sennett (2008) has embarked on an effort to reclaim making. His belief is that "people can learn about themselves through the things they make" (8). In short, he argues, material culture matters. Sennett is part of a burgeoning chorus of scholars and activists who have argued for a revival of at least an appreciation of making if not actually applying ourselves as a nation to making.

In 2005, O'Reilly, a publisher of technical books, launched *Make* magazine. Around the same time, the Discovery Channel television network premiered *Dirty Jobs*, a show focused on the various kinds of jobs that, although manual in nature and often dirty and smelly—the show seems to relish disgorged guts and "poo" as its host Mike Rowe is fond of calling at least one bodily function—are also often revealed to be an important part of the complex web of actions that underwrite modernity. The Travel Channel's *Made in America* was more in keeping with the older ways of doing things, but shows that featured guys welding, such as *Orange County Choppers*, presaged what are now dozens of similar programs featuring guys in shops working on cars, guns, motorcycles, and assorted other equipment and objects.

It's as if the strong wind that carried the United States through the nineteenth century and through much of the twentieth has suddenly slackened and we have lost, quite literally, our driving force.

18. Hymes's study was originally published as "Jeffers's Artistry of the Line" in 1991 in *Centennial Essays for Robinson Jeffers*, and it was reprinted in a 2003 collection of his essays, *Now I Know Only So Far: Essays in Ethnopoetics* (Lincoln: University of Nebraska Press) with the poet's complete name in the title of the essay. (Context is everything.)

19. This binding of people to the land was a foundational component of romantic nationalism, a part of the intellectual history of folklore studies with which more folklorists are familiar. Richard Bauman and Charles Briggs offer the most thoroughgoing history I have yet seen, with regard to folklore studies, of the invention of the folk and their place in the world in *Voices of Modernity* (2003).

20. "When large continental ice sheets covered the Midwestern United States, summer and spring melting at their southern edges created huge volumes of meltwater that flooded down the Mississippi, Missouri, and Ohio Rivers. As the ice sheet melted during the spring and summer, the meltwater carried large quantities of glacial sediment downstream with it. This sediment included considerable silt-size particles created by the grinding of ice sheets over bedrock and silt derived from Late Pleistocene sand dunes in Nebraska and eastern Colorado. The meltwater flowing down an extensive braided stream system spread the glacial sediment, including large volumes of silt, over the Pleistocene floodplain of the Mississippi River"; Heinrich (2008), 5.

21. Fred Kniffen first discerned the potential for a French cultural footprint in his early work on Louisiana house types (1936), and the idea developed over a number of years and publications and is perhaps best observed in his 1968 survey of Louisiana geographies, *Louisiana: Its Land and People* (1968). Many of the subsequent maps of Louisiana's cultural regions are based on Kniffen's pioneering work. (For examples of such work, see the fine collection of maps on the Louisiana Folklife website: http://www.louisianafolklife.org/LT/creole_maps.html.)

22. Geologists call this processing of deposited materials compacting *subsidence*. It is this process, land built up by centuries of flooding slowly settling down, that actually explains a lot of what is called *coastal erosion*. No doubt, the loss of vegetation caused by canals cut into the landscape have hastened the loss of a lot of habitat, but the bulk of the land lost is to subsidence.

23. I have seen many such trailers pulled around by farmers. I have even helped place large, heavy metal objects on top of plenty of trailers, wondering all the while if the trailer would make it to the end of the drive, let alone its final destination. Trailers are trailers in this world. This one was no different. As long as the wheels rolled and kept their cargo off the ground, the trailer would be pulled to its next stop.

24. Babineaux (1967).

25. Dethloff (1988), 372.

26. Babineaux (1967) noted that "the Southern Pacific Railroad financed the printing of numerous pamphlets and books which publicized the Gulf Coast. Among these were *The Gulf Coast of Louisiana: "Where Nature Smiles"*; *Vermilion Parish Louisiana: The Farmer's Road to Wealth*; *The Appeal of Louisiana to the Western Farmer*; and *Southwest Louisiana along the Line of the Southern Pacific*. The literature encouraged immigrants to "settle in the prairie region and begin a life of successful farming."

27. Dethloff (1988), 374.

28. Ibid., 375.

29. These figures are from the United States Bureau of the Census, United States Census (Washington, DC: Department of Agriculture), 3:759, as noted in Dethloff (1988).

30. Lawson Babineaux's 1967 thesis, "A History of the Rice Industry of Southwestern Louisiana," remains one of the central references for people interested in the early economics of rice farming.

31. Wilson's speech was quoted in the Southern Pacific Railroad Company's *Southwest Louisiana Up to Date* (81). He was speaking before a meeting of the Louisiana Agricultural Society. This passage from Wilson's speech is in Babineaux's (1967) chapter on "The Development of Commercial Rice Farming in Southwestern Louisiana."

32. T. S. Adams, "Report to the Governor of Louisiana," *Biennial Report of the Louisiana Commissioner of Agriculture* (April 1892), 6.

33. Perhaps as important as the mechanization was the introduction and diffusion of the idea of forming companies in order to achieve larger goals. One of the first instances of this, after the land companies themselves, was the creation of an alternative system of mills as a response to the oppressive pricing of the New Orleans mills—which found themselves obsolete within a decade. Another instance, this time in response to the droughts of the mid-1890s, was the development of a number of canal companies. Canals acted not only as means of conveying water

from a source, like a bayou (usually by pumping), but also as reservoirs, holding water until it was needed. Some canal companies offered more than water: they would give farmers land and seed as well in return for a share of the crop. An advertisement circa 1900 by the Vermilion Development Company read: "To any party having working-stock we will build a house and pasture. Any amount of land required will be furnished. . . . The seed required will be advanced, same to be returned after harvest. A complete pumping outfit will be rented at cost for the purpose of irrigating the rice field. We pay our share of threshing and furnish our share of sacks. We ask as our share one-fourth of the total crop." Such a successful engine—technology—and economic cooperation—for inputs like water as well as outputs in terms of milling—drove land values from fifty cents to ten dollars within a decade, which in turn drove farmers into eastern Texas and, later, southern Arkansas, developing those areas as rice producers as well.

34. The term *bull wheel* is a later appellation, used first in the oil industry and later applied to all such mechanisms. The first bull wheels, however, appeared on farm implements: Cyrus McCormick's 1834 reaper featured a bull wheel, which remained an important part of agricultural machinery until small gasoline engines became more widely available in the 1920s.

35. Acadia Parish lies at the heart of Louisiana rice country, usually producing the most barrels, but it is followed closely by the neighboring parishes of Evangeline (to the north), Jefferson Davis (to the west), and Vermilion (to the south). Radiating outward, and producing less rice, are Saint Landry, Saint Martin, Cameron, Beauregard, Rapides, and Avoyelles Parishes. Concordia Parish, situated as it is along the river, bridges the coastal prairies to the south with the river delta parishes and counties to the north. (Based on information from the US Department of Agriculture's National Agricultural Statistical Service, 2010.)

36. The total amount of land planted in rice in Louisiana has been a little over 400,000 acres for the past few years, down from 500,000 to 600,000 acres twenty years ago. In that span of time, however, the amount of rice each acre produces has climbed from an average of thirty or so barrels per acre to averages in the mid-forties. The result has been some seventeen million barrels of rice produced on average during the past decade. It should be noted that the USDA records everything in terms of hundredweights, and so the numbers for these periods are actually a rise from an average of forty-five hundredweights per acre to hundredweights regularly in the seventies, with 2013 peaking at eighty-one hundredweights per acre. Farmers, however, use the older measurement of a barrel, which represents roughly 162 pounds of rice. In order to keep the numbers they report to me each year in sync

with the figures provided by the USDA, I have converted hundredweights to barrels by simply multiplying by 0.6173. In doing so, I am aware that I am completely glossing over matters of raw versus milled rice, but once you add in that I have probably blended or confused acres planted with acres harvested at some point in compiling all these figures, then the reader simply needs to remember that this history is meant largely to be suggestive of the larger complexities and not nearly the accurate representation that it truly deserves.

37. This tale was recovered by George Reinecke, who published both the original French version as well as the English translation used here in the 1994 volume of *Louisiana Folklore Miscellany*. Readers interested in these tales should consult Reinecke's article for his excellent coverage of the matter.

38. Folklorists in particular will recognize that it is a version of tale type 513B, a member of the class of tales called *magic tales* in the ATU (Aarne-Thompson-Uther) classification system. This version is a member of the subclass of supernatural helpers, with 513A being the helpers without the ship. This places it firmly in the middle of the ordinary folktales (types 300–1199), within the category of *tales of magic*. Motifs present in the narrative include L13, a reversal of fortune that focuses on the compassionate youngest son, as well as Q40, kindness rewarded.

39. The version that sits on my shelf, *The Fool of the World and the Flying Ship*, was given to me by my godmother, Sharon Rickey, when I was only three. Illustrated by Uri Shulevitz, the text is from Arthur Ransome's *Old Peter's Russian Tales*, originally published in 1916. The tale previously appeared in Andrew Lang's *Yellow Fairy Book* in 1894. Another Russian version of the tale was published by Alexander Afanasyev as part of his eight-volume *Russian Fairy Tales* series (1855–63), which he modeled directly on the efforts of the Brothers Grimm. (That I keep this childhood book on my shelves probably reveals that it was inevitable that I write about these boats.)

40. In *Swapping Stories*, the book from which this tale is taken, Lindahl, Owens, and Harvison (1997) set out to capture a much wider variety of genres and topics of vernacular discourse than is usually seen, especially in regions like Louisiana, which get renown for a particular thing and soon much else is lost from view.

41. The story of a pirate in a tree is one I recorded myself. It was told to me by Oscar Babineaux of Rayne, as we sat in the cool of his home one July afternoon. It is one of several legends I examine in a discussion of the performance that took place that afternoon (Laudun 2012).

42. Brassieur made this observation during an interview in his campus office one afternoon in 2007.

43. Knipmeyer's essay was actually the last chapter of his dissertation, "Settle-

ment Succession in Eastern French Louisiana," at Louisiana State University (1956), which he wrote under the direction of Fred Kniffen. Kniffen must have mentioned the dissertation, or at least the chapter, to Henry Glassie, whom he unofficially tutored while Glassie was a student at Tulane University in New Orleans. Glassie regularly made the journey to LSU in Baton Rouge to study with Kniffen, who was one of the first scholars to recognize the larger pattern in folk housing that he called the "I-house." Glassie cited Knipmeyer's dissertation in *Pattern in the Material Folk Culture of the Eastern United States* (1968), a text with a parallel title to Knipmeyer's own dissertation, and later edited the last chapter, which was on folk boats, for inclusion in Don Yoder's *American Folklife* (Knipmeyer 1976). Malcolm Comeaux's essay appears in *Louisiana Folklife* (1985).

44. Knipmeyer quotes extensively from Le Page du Pratz's description of the manufacture of a pirogue. I quote it here for those interested, since the Yoder collection is increasingly difficult to find and many will not have access to Le Page du Pratz in the original:

> They always cut a tree close to the ground so that the fire they built at the foot of the tree would more easily consume the filaments and fibers of the wood which the axe had mashed. Finally, with much trouble and patience, they managed to bring the tree down. This was a long piece of work, so that in those times they were much busier than at present, when they have the axes we sell them. From this it happens that they no longer cut a tree down at the base, but at the height which is most convenient.
>
> This occasions them an infinite amount of labor, since they have no other utensils in this work than wood for making fire and wood for scraping, and only small wood is required to burn. In order to set fire to this tree destined for making a pirogue, a pad of clay, which is found everywhere, has to be made for the two sides and each end. These pads prevent the fire from passing beyond and burning the sides of the boat. A great fire is made above, and when the wood is consumed it is scraped so that the insides may catch fire better and may be hollowed out more easily, and they continue thus until the fire has consumed all the wood in the inside of the tree. And, if the fire burns into the sides they put mud there which prevents it from working farther than is demanded. This precaution is taken until the pirogue is deep enough. The outside is made in the same manner and with the same attention.
>
> The bow of the pirogue is made sloping, like those of the boats which one sees on the French rivers. This bow is as broad as the body of the pirogue. I have seen some 40 feet long by 3 feet broad. They are about 3 inches thick which makes them very heavy. These pirogues can carry 12 persons and are all of buoyant wood. (Knipmeyer 1976, 108–9; Le Page du Pratz 1774 [1947], 66–67)

45. Interestingly, in Puerto Rico *pirogue* refers to a shaved ice dessert that is, at least in Louisiana, known as a *snow cone*. At least one author has speculated that the name is a portmanteau of the Spanish words for pyramid and water: *piramide*

plus *agua*. It is, perhaps, just as likely that it is a reference to the shavings created during the creation of a dugout pirogue. At the very least, this historical dimension of the contemporary word may have helped solidify its current usage. (Leonor Toro, "Luisito and the Piragua," p. 12, New Haven, CT: New Haven Migratory Children's Program, Hamden–New Haven Cooperative Education Center; ERIC #ED209026; retrieved July 14, 2008).

46. Perhaps no better example of the current use of the chaland form can be found than in the crawfish push boat, which is usually only six feet long and three feet wide with a raised bar in the back so that the boat can be pushed, by hand, ahead of its operator, who walks the line of traps. They are small and simple in design and manufacture and have reemerged as an alternative to the mechanized boats, especially in the wake of the 2008–9 spike in fuel prices.

47. Comeaux's argument is based, in part, on the work of John Amos Johnson, whose 1963 dissertation is titled "Pre-Steamboat Navigation on the Lower Mississippi" (Louisiana State University).

48. It should be noted that both *joe* and *john*, as well as *jon*, *boats* can appear as one word: *joeboat*, *johnboat*. I have never seen anyone venture an explanation for the use of those particular names and not others, only that jon/john boats are the common term among English speakers in the United States. As Dana Everts-Boehmn (1991) notes, boats of this form are known as *punts* or *scows* in the British Isles.

49. Guirard and Brassieur (2007) note that many of the fish-buying boats were themselves scaled-up versions of putt-putts, some of them more than twenty-six feet long and over six feet wide (55).

50. Older residents will remember that 1964 was the year that Hilda struck. Of the twelve storms that year, only one struck Louisiana. Hilda was one of only four storms to develop in the Caribbean that year, and only one of two to make US landfall. The other storm never reached hurricane strength.

51. It's not clear if Songe had any direct experience of long-tail boats, but the principle of a swivel-mounted outboard engine that produces both vectored thrust and has the ability to switch quickly into neutral (by dropping the handle and rocking the propeller up into the air) is clearly parallel to developments in southeast Asia.

52. Many historians of technology now believe that much of the agricultural revolution is due to something known as the *medieval climactic optimum,* a period during the Middle Ages when temperatures were warmer than normal, making growing seasons throughout Europe longer, and thus making it possible for some

crops to be grown in northern Europe for the first time. The mechanical innovations of the plow and horse collar may have sprung from the expanded possibilities for agriculture during this period.

53. For a more detailed account of the construction and use of pillow traps, see Shirley and Lutz (2009).

54. Dwight Landreneau worked for the LSU AgCenter for twenty years as a 4-H agent, county agent, and area crawfish specialist. He was, quite literally, in the field during the time that the crawfish boat took form.

55. I have heard the cylinder referred to as the *throat,* akin to the throat of a combine; as a *spout,* akin to a bottle; and as a *chimney,* which is the same word often used to describe the tall, muddy tubes crawfish make when they burrow on land.

56. The research station between Rayne and Crowley—sometimes known as the rice station, although it also addresses soybean and crawfish matters because those are the two other crops most popular among rice farmers—hosts an annual field day that almost every farmer in the area has been to at some time or another. The relationship between the station and the area farmers is quite warm: more than one farmer has spoken with genuine affection and gratitude for the efforts of the station's staff and scientists. The agricultural research station has been a highly successful experiment in allowing a direct connection between various kinds of research and the practitioners who have the most interest, and investment, in that science, and it leads to a lot of conversations about the nature of science. Sadly, the research station system has been affected by deep cuts to higher education in the state in the past five years or so.

57. Benoit initially built the boats for $1,200 to $1,500, but the hulls alone, which he bought from Sears, cost about $700.

58. Guillory would later appear in a CBS News report on the rise of crawfish agriculture. A portion of that video, with a brief glimpse of one of Benoit's boats as Guillory operates it, is available at http://www.flickr.com/photos/johnlaudun/3694274037/in/photostream/.

59. The description that follows may be too detailed for some, but I wanted to capture the nuances of at least one builder. We have amazingly detailed descriptions of how words get assembled into sentences, sentences into paragraphs, paragraphs into chapters, and chapters into books, but we have frightfully few accounts of how metal gets welded to metal to form the objects upon which our lives so often depend. I have taken as my inspiration, and rationale, for doing so the work of Charles Zug on North Carolina potters and of Douglas Harper in his examination of a single man at work in his repair shop.

60. Although the farmers recognize that watersheds are shared and that the landscape on which they work is part of something larger, be it "the Cove" or "Cajun country" or simply "Louisiana," they largely discuss it in terms of established pieces: this family's land versus that family's. Even if two farms share, for example, a high point or a canal, I have never heard such things referenced.

61. How technology is figured in discourse, as both an idea and ideal as well as a practice, is the subject that Marx took up repeatedly. Toward the end of *The Machine in the Garden* (1964), he noted that "the sudden appearance of the machine in the garden is an arresting, endlessly evocative image. It causes the instantaneous clash of opposed states of mind: a strong urge to believe in the rural myth along with an awareness of industrialization as counterforce to the myth. . . . It appears everywhere in American writing" (229).

62. And then there are people like Frank Verret, an older man living in Saint Mary Parish whom I once interviewed about his life as a worker in the cane fields. Wanting to know how to reach him if I had some additional questions, Verret directed me to call his daughter; she had a telephone and lived next door, he said. She would tell him I had called or go get him. It turns out that Mr. Verret, in 2000, saw no need to own a telephone; everyone he knew lived nearby and if he wanted to talk with them, he would drive to their house.

63. The definitions within which the work gets done are not necessarily the problem. The American Folklife Center of the Library of Congress, for example, begins its definition of *material culture* with a list of things: "Houses, barns, and other traditional buildings constitute a subcategory of material culture known as vernacular architecture. Other objects of interest include baskets, boats, clothing, furniture, metalwork, pottery, and quilts." It's not a bad list, and one supposes that metalwork might encompass fabrication and repair work. The summary that follows helps: "In general, folklore studies of material culture have favored handmade objects such as these, and craftsmanship itself has been a special focus." So there is room within folklore studies, itself often the fuzzy fringe of the humanities, for a wider view of things and their makers. The entry on material culture is part of an "illustrated guide" published on the American Folklife Center's website (www.loc.gov/folklife/guide/materialculture.html). The images that accompany the entry on material culture are of a quilt, a cemetery memorial, a wooden boat, a hammock, and other such forms as well as junkyard robot sculptures.

64. *The Wheelwright's Shop* (1930), 73–74. Sturt is especially eloquent on the matter of repair work and how demanding it is: "From repairs, in fact, came the teaching which kept the wheelwrights' art strongly alive. A lad might learn from

older workmen all about the tradition—all that antiquity had to teach—but at repairs he found out what was needful for the current day; what this road required, and that hill; what would satisfy Farmer So-and-So's temper, or suit his pocket; what the farmer's carter favored or his team wanted. While 'new-work' was largely controlled by proven theories and by well-tried fashions, on the other hand repairs called for ingenuity, adaptiveness, readiness to make shift. It wasn't quite enough to know how to do this or that; you needed also to know something about why, and to be ready to think of alternative dodges for improvising a temporary effect, if for any reason the time-honoured methods known to an apprentice could not be adopted" (176).

65. The phrase *indiscriminate bolus* is taken from T. S. Eliot's essay on tradition and occurs when he encourages artists working within a tradition neither to ignore nor be overwhelmed by what has come before ("Tradition and the Individual Talent").

66. Sturt (1930), 75.

67. In the past decade or so, as the amount of visual and verbal information available to us has grown considerably, there has been increasing interest in alternative interfaces for operators, be they drivers of cars or deep-sea divers. Beginning with the conclusion that individuals are otherwise overloaded with visual and verbal information, the research has mostly focused on how adding various kinds of haptic devices to deliver vibrotactile cues can aid drivers or divers in processing diverse flows of information. For many, the ability to add a vibrotactile device to a car or to a watch is a way to overcome the visual overload many operators navigating complex environments feel. See, for example, the work by Rizzolatti and Fadiga (1997) as well as the work of Ho et al. (2005).

68. Frake (1985), 256.

69. Frake offers his readers a brief refresher into the relationship between the two kinds of time that sailors once tracked simultaneously and in relationship to each other: lunar and solar. Lunar time dominated sailing concerns during the Middle Ages precisely because sailing along and between coasts was the paramount concern, and would continue to be a significant concern even as sailors ventured more widely, because their successful return to a port would be determined by their ability to navigate its channels and shoals. Much of this navigation is a function of the tides, and the foundation of the tides is the gravitational pull of the moon on Earth's oceans. Even today, most readers are aware that high tide occurs when the moon is directly overhead, which can also be described as *lunar noon,* and that it also occurs when the moon is on the opposite side of Earth, in what is known as

lunar midnight. Because the moon circumnavigates Earth in one solar day, there are two high tides and two low tides, but because a lunar orbit is actually somewhat longer than a day, forty-eight minutes longer, the tides are about six hours and twelve minutes apart. This unit of lunar time is known among sailors and people who live along a coast simply as a *tide.*

What is important about a tide is that it frequently determines the accessibility of a port, with some ports only being navigable at particular tidal times, often somewhere near high tide. Remarkably, Frake notes, "medieval sailors pressed their cognitive map of directions, the compass rose, into service as a schema for representing and manipulating temporal information" (264). The compass rose was, for these sailors, both a means of determining direction as well as a means for calculating time. The rose is made up of thirty points: eight full points, eight half points, and sixteen quarter points. Used directionally, the full points represent the cardinal directions of north, east, south, and west, as well as the four ordinal points of northeast, southeast, southwest, and northwest. The half points represent the bisection of the full points, such as east-northeast, usually noted as ENE, and the quarter points represent a bisection of the full and half points, usually noted with a *by,* such as northeast by east (NEbE). Imagined as a twenty-four-hour clock representing the relationship of astronomical bodies to an observer at a particular location on Earth, each of the full points represents the hours of midnight, 3:00 a.m., 6:00 a.m., 9:00 a.m., noon, 3:00 p.m., 6:00 p.m., and 9:00 p.m. The half points mark the half hours of 1:30, 4:30, 7:30, and 10:30; and the quarter points mark the passage of forty-five minutes. Three hours, one-and-a-half hours, and three-quarters of an hour seem useless until we remember that the tide runs approximately every six hours.

Like the sun, the moon bears due south for observers in the northern hemisphere, and thus another name for lunar noon is *moon bears south.* Conversely, *moon bears north* describes, in effect, lunar midnight, the moment when the moon is on the opposite side of Earth from the observer. Thinking of the moon's position in this way allows one to describe the moon as being WSW, or six compass points past south. Because each compass point equals forty-five minutes, a WSW moon occurs four and a half hours after lunar noon, which would be high tide. Such a framework allowed medieval sailors to compress a high amount of navigational information into statements like "all havens be full at WSW moon between the Start and the Lizard" (Taylor 1956, 132). Given that the tides shift by about a compass point, or forty-five minutes, every day, a sailor could then take such a known fact and, using the compass rose, calculate what he needed to know in relationship to his own moment in time. As Frake notes, "If WSW moon corresponds to 4:30 (we can now ignore am and pm) at full moon, and it is now five days past full moon, we can count five compass points past WSW to NW by W, a point which marks the solar time of 8:15" (265). High tide will occur at 8:15 in the morning and evening five days after the full moon locally. Just as important for the sailor was the ability to know, quickly, when the half and quarter tides before and after the high and low tides were, since they would indicate whether one was facing an ebb or flood tide. (Tides are considerably more complex than I am representing them here. For one, although water appears simply to rise and fall at a particular point, it is in fact part of a giant tidal wave that circumnavigates the globe in pursuit of the moon. The ever-westward course of this wave interacts with oceanic currents as well as the topographies of the seafloor and coastlines to produce a wide variety of local conditions; see Waters [1976], 421–22.)

70. Frake (1985), 266.

71. Hutchins (1995), 361, 362. Later in the conclusion of *Cognition in the Wild*, Hutchins argues that much of the work done in artificial intelligence and in cognitive psychology consistently focuses on sociocultural systems but mistakes them for individual minds.

72. It is not entirely clear who started making hulls—although Harold Benoit noted that one of the reasons he got out of the business was because people were moving to heavier boats. The boats he encountered were made by Gerard Olinger.

73. Author Kevin Kelley likes to switch the point of view and ask "what does technology want?" in order to explore ideas suggested by Stuart Kauffman's notion of the adjacent possible.

74. The sorting of crawfish by size is not something treated here. At least three sizes are observed with regularity locally: select, peeler, and everything else. Select crawfish usually command a premium price at local restaurants or they are shipped to markets, such as Houston or Dallas, whose diners/purchasers are not aware that crawfish come any smaller. Peelers are crawfish normally boiled or parboiled by a processor and placed into one-pound packages for use in stews, étouffées, and other dishes. Everything else is what you get when you order boiled crawfish at a restaurant or pick up a sack of crawfish to boil at home. The exact size for what constitutes any of these three categories is subject to debate, especially when you feel you have gotten a platter of peelers. Some crawfish farmers and processors have tried to introduce additional sizes, but none has emerged as more useful than the current vernacular scheme.

75. The official language, from the 2012 regular session, is that House Bill 746,

with a host of signatories, enacts "R.S. 49:170.17, relative to state symbols; to provide for the official state boat; and to provide for related matters." The legal act reads as follows: "Be it enacted by the Legislature of Louisiana: Section 1. R.S. 49:170.17 is hereby enacted to read as follows: §170.17. State boat: There shall be an official state boat, which shall be the pirogue. The use of its likeness or image on official documents of the state and with the insignia of the state is hereby authorized."

76. The Acadian Memorial website notes that "the Wooden Boat Congrès . . . began as a means of honoring and displaying traditional South Louisiana vessels, especially those indigenous to the Atchafalaya Swamp region like pirogues, dugout canoes and chalons, known locally as 'putt-putt' boats. Each year boat enthusiasts come from all over Louisiana and even other states to display their antique boats. A Re-enactment of the Arrival of the Acadians is always a part of this festival, with costumed Acadians paddling pirogues." (http://acadianmemorial.org/)

77. *Louisiana Story* (1948) is regularly misunderstood as being a documentary, when, in reality, it was a fictional, feature-length film sponsored by the Standard Oil Company. It presents a Cajun family, made up of three strangers in reality, who become wealthy thanks to the discovery of oil on their property in the swamp. It may very well be the founding cinematic document that makes Cajuns out to be swamp dwellers.

78. In their discussion of both Henry Schoolcraft and Franz Boas in *Voices of Modernity*, Richard Bauman and Charles Briggs (2003) offer a lucid history of the "bridge" dynamic, wherein the folklorist or anthropologist often works very closely with someone who is both within the studied community and also somewhat outside of it. Both Schoolcraft and Boas were quick, it would seem, to smooth over the possible mixed, or hybrid, nature of the indigenous translator in order to make their own work more authentic. This dynamic remains with folklore studies to this day; it is, of course, prominent in Louisiana studies with "natives," of which I am one, I suppose, somehow having a better relationship with some indigenous reality, whatever that may be.

79. Some formalists, like proponents of artificial intelligence, maintain that the kind of "sloppiness" or "slipperiness" of lived reality is simply levels of complexity yet to be mapped and formalized. Such an argument is well beyond the scope of this work. Certainly, some researchers in this area have their doubts. See Hubert Dreyfus and Stuart Dreyfus's *Mind over Machine* (1986).

BIBLIOGRAPHY

Andrews, Francis B. 1922. *The Mediæval Builder and His Methods*. Transactions and Proceedings of the Birmingham Archæological Society XLVIII (Birmingham, England). Repr. 1999, Mineola, NY: Dover.

Ancelet, Barry Jean. 1994. *Cajun and Creole Folktales: The French Oral Tradition of South Louisiana*. New York: Garland.

Babineaux, Lawson P., Jr. 1967. "A History of the Rice Industry of Southwestern Louisiana." MA thesis, Lafayette: University of Southwestern Louisiana. http://ereserves.mcneese.edu/depts/archive/FTBooks/babineaux.htm

Barth, Fredrik. 1969. "Introduction." In *Ethnic Groups and Boundaries: The Social Organization of Cultural Difference*, edited by Fredrik Barth, 9–38. New York: Little, Brown.

Bauman, Richard, and Charles Briggs. 2003. *Voices of Modernity: Language Ideologies and the Politics of Inequality*. Studies in the Social and Cultural Foundations of Language. Cambridge: Cambridge University Press.

Baxandall, Michael. 1985. *Patterns of Intention: On the Historical Explanations of Pictures*. New Haven, CT: Yale University Press.

Bogatyrev, Petr, and Roman Jakobson. 1978. "On the Boundary between Studies of Folklore and Literature." In *Readings in Russian Poetics: Formalist and Structuralist Views*, edited by Ladislav Matejka and Krystina Pomorska, 91–93. Ann Arbor, MI: Slavir.

Bourne, Joel K., Jr. 2007. "New Orleans: A Perilous Future." *National Geographic*, August, 32–67. http://ngm.nationalgeographic.com/2007/08/new-orleans/new-orleans-text.

Brady, Erika. 1990. "Mankin's Thumb on Nature's Scale: Trapping and Regional Identity in the Missouri Ozarks." In *Sense of Place: American Regional Cultures*, edited by Barbara Allen and Thomas Schlereth, 58–73. Lexington: University Press of Kentucky.

Brasseaux, Carl. 1987. *The Founding of New Acadia: The Beginnings of Acadian Life in Louisiana, 1765–1803*. Baton Rouge: Louisiana State University Press.

————. 1992. *Acadian to Cajun: Transformation of a People, 1803–1877*. Jackson: University Press of Mississippi.

Comeaux, Carmen, Janet Schexnayder Elias, and Subrata Dasgupta. 2006. "Creativity, Cognition, and the Case Study Method." In *Frontiers in Cognitive Psychology*, edited by Michael A. Vanchevsky, 105–25. Hauppauge, NY: Nova Science.

Comeaux, Malcolm. 1985. "Folk Boats of Louisiana." In *Louisiana Folklife: A Guide to the State*, edited by Nicholas R. Spitzer, 160–78. Baton Rouge: Louisiana Folklife Program/Louisiana Department of Culture, Recreation, and Tourism.

Dethloff, Henry C. 1988. *A History of the American Rice Industry, 1685–1985*. College Station: Texas A&M University Press.

Dreyfus, Hubert, and Stuart E. Dreyfus. 1986. *Mind over Machine The Power of Human Intuition and Expertise in the Era of the Computer*. New York: Free Press.

Edgerton, David. 2007. "Creole Technologies and Global Histories: Rethinking How Things Travel in Space and Time." *Journal of the History of Technology* 1: 75–112.

Everts-Boehmn, Dana. 1991. "The Ozark Johnboat: Its History, Form, and Functions." In *The Masters and Their Traditional Arts*, edited by Ray Brassieur and Howard Wight Marshall. Columbia: Missouri Arts Council/University of Missouri Cultural Heritage Center.

Ferguson, Eugene. 1992. *Engineering and the Mind's Eye*. Cambridge, MA: MIT Press.

Frake, Charles O. 1985. "Cognitive Maps of Time and Tide among Medieval Seafarers." *Man*, n.s., 20 (2) (June 1): 254–70.

Gilmore, Janet C. 1999. *The World of the Oregon Fishboat: A Study in Maritime Folklife*. Pullman: Washington State University Press.

Glassie, Henry. 1968. *Pattern in the Material Folk Culture of the Eastern United States*. Philadelphia: University of Pennsylvania Press.

————. 1975. *Folk Housing in Middle Virginia: A Structural Analysis of Historic Artifacts*. Knoxville: University of Tennessee Press.

————. 1982. *Passing the Time in Ballymenone: Culture and History of an Ulster Community*. Philadelphia: University of Pennsylvania Press.

————. 1993. *Turkish Traditional Art Today*. Bloomington: Indiana University Press.

————. 1999. *Material Culture*. Bloomington: Indiana University Press.

Gomez, Gay. 1998. *A Wetland Biography: Seasons on Louisiana's Chenier Plain*. Austin: University of Texas Press.

Guirard, Greg, and C. Ray Brassieur. 2007. *Inherit the Atchafalaya*. Lafayette: Center for Louisiana Studies/University of Louisiana.

Harper, Douglas. 1987. *Working Knowledge: Skill and Community in a Small Shop*. Chicago: University of Chicago Press.

Heinrich, Paul V. 2008. "Loess Map of Louisiana." Public Information Series 12. Baton Rouge: Louisiana Geological Survey.

Ho, Cristy, Hong Z. Tan, and Charles Spence. 2005. "Using Spatial Vibrotactile Cues to Direct Visual Attention in Driving Scenes." *Transportation Research Part F: Traffic Psychology and Behaviour* 8 (6): 397–412. doi:10.1016/j.trf.2005.05.002.

Hufford, Mary. 1992. *Chaseworld: Foxhunting and Storytelling in New Jersey's Pine Barrens*. Philadelphia: University of Pennsylvania Press.

Hutchins, Edwin. 1995. *Cognition in the Wild*. Cambridge, MA: MIT Press.

Hymes, Dell. 1991. "Jeffers's Artistry of the Line." In *Centennial Essays for Robinson Jeffers*, edited by Robert Zaller, 226–47. Cranbury, NJ: Associated University Presses.

Jakobson, Roman, and Petr Bogatyrev. 1929. "On the Boundary between Studies of Folklore and Literature." In *Readings in Russian Poetics*, 91–93. Tr. Herbert Eagle. Ed. Ladislav Matejka and Kystyna Pomorska. Ann Arbor: Michigan Slavic Publications, 1978.

James, John. 1982. *The Master Masons of Chartres*. London: Routledge/Kegan Paul.

Johnson, John Amos. 1963. "Pre-Steamboat Navigation on the Lower Mississippi." PhD diss., Baton Rouge: Louisiana State University.

Jordan, Rosan Augusta, and Frank de Caro. 1996. "'In This Folk-Lore Land': Race, Class, Identity, and Folklore Studies in Louisiana." *Journal of American Folklore* 109 (431): 31–59.

Kauffman, Stuart A. 2003. "The Adjacent Possible: A Talk with Stuart A. Kauffman". *Edge.com*. Edge Foundation. http://www.edge.org/conversation/the-adjacent-possible.

Kelly, Kevin. 2011. *What Technology Wants*. New York: Penguin Books.

Kniffen, Fred. 1936. "Louisiana House Types." *Annals of the Association of American Geographers* 26 (4): 179–93. doi:10.2307/2569532.

———. 1968. *Louisiana: Its Land and People*. Baton Rouge: Louisiana State University Press.

Knipmeyer, William. 1976. "Folk Boats of Eastern French Louisiana," edited by Henry Glassie. In *American Folklife*, edited by Don Yoder, 105–50. Austin: University of Texas Press.

Laudun, John. 2012. "'Talking Shit' in Rayne." *Journal of American Folklore* 125 (497): 304–26.

Le Page du Pratz, Antoine-Simon. 1774 (1947). *History of Louisiana*. Repr., New Orleans: Pelican Press.

Lindahl, Carl, Maida Owens, and C. Renée Harvison, eds. 1997. "The Alligator Peach

Tree." In *Swapping Stories: Folktales from Louisiana*, 218–19. Jackson/Baton Rouge: University Press of Mississippi/Louisiana Division of the Arts.

Marx, Leo. 1964. *The Machine in the Garden: Technology and the Pastoral Ideal in America.* New York: Oxford University Press.

Post, Lauren. 1990. *Cajun Sketches: From the Prairies of Southwest Louisiana.* Baton Rouge: Louisiana State University Press.

Reinecke, George. 1994. "A Louisiana Black Creole Version of 'The Land and Water Ship.'" *Louisiana Folklore Miscellany* 9:19–29.

Rizzolatti, Giacomo, and Luciano Fadiga. 1997. "The Space around Us." *Science* 277 (5323) (July 11): 190. doi:10.1126/science.277.5323.190.

Robin, Charles-César. 1966. *Voyage to Louisiana, 1803–1805.* Translated by Stuart O. Landry. New York: Firebird Press.

Saucier, Corrine. 1962. *Folk Tales from French Louisiana.* Baton Rouge: Louisiana State University/Claitor's.

Sennett, Richard. 2008. *The Craftsman.* New Haven, CT: Yale University Press.

Shirley, Mark, and C. Greg Lutz. 2009. *Crawfish Trap Design and Construction.* SRAC Publication No. 2404. Stoneville, MS: Southern Regional Aquaculture Center.

Sturt, George. 1923. *The Wheelwright's Shop.* Cambridge: Cambridge University Press.

Taylor, Eva G. R. 1956. *The Haven-Finding Art: A History of Navigation from Odysseus to Captain Cook.* London: Hollis and Carter.

Vlach, John Michael. 1981. *Charleston Blacksmith: The Work of Philip Simmons.* Athens: University of Georgia Press.

Waters, David W. 1976. *The Rutters of the Sea: The Sailing Direction of Pierre Garcie, a Study of the First English and French Printed Sailing Directions.* New Haven, CT: Yale University Press.

Zug, Charles G., III. 1986. *Turners and Burners: The Folk Potters of North Carolina.* Chapel Hill: University of North Carolina Press.

INDEX

Page numbers in *italics* indicate an illustration.